Second Edition

Practical Design Calculations for Groundwater and Soil Remediation

Second Edition

Practical Design Calculations for Groundwater and Soil Remediation

Jeff Kuo

CRC Press
Taylor & Francis Group
Boca Raton London New York

CRC Press is an imprint of the
Taylor & Francis Group, an **informa** business

CRC Press
Taylor & Francis Group
6000 Broken Sound Parkway NW, Suite 300
Boca Raton, FL 33487-2742

© 2014 by Taylor & Francis Group, LLC
CRC Press is an imprint of Taylor & Francis Group, an Informa business

No claim to original U.S. Government works

Printed on acid-free paper
Version Date: 20140311

International Standard Book Number-13: 978-1-4665-8523-2 (Hardback)

Library of Congress Cataloging-in-Publication Data

Kuo, Jeff.
    Practical design calculations for groundwater and soil remediation / author, Jeff Kuo. -- Second edition.
        pages cm
    Summary: "This new edition has updates throughout the text, and it presents all new material on air sparging, dual phase extraction, chemical oxidation, and vacuum pumps, as well as several new practical examples. It illustrates the engineering calculations needed during site assessment and remedial investigations. It also shows readers how to estimate the rates of groundwater movement and plume migration. This book provides important design calculations for vadose zone soil remediation as well as groundwater remediation"-- Provided by publisher.
    Includes bibliographical references and index.
    ISBN 978-1-4665-8523-2 (hardback)
    1. Soil remediation--Mathematics--Problems, exercises, etc. 2. Groundwater--Purification--Mathematics--Problems, exercises, etc. 3. Engineering mathematics--Formulae. I. Title.

TD878.K86 2014
628.1'68--dc23   2014007195

Visit the Taylor & Francis Web site at
http://www.taylorandfrancis.com

and the CRC Press Web site at
http://www.crcpress.com

To my wife Kathy, daughters Emily and Whitney, and my mom

Contents

Preface

The focus of the hazardous waste management business has switched from litigation and site assessment to remediation. Site restoration usually proceeds through several phases and requires a concerted, multidisciplinary effort. Thus, remediation professionals come from a variety of technical and educational backgrounds, including geology, hydrology, chemistry, microbiology, meteorology, toxicology, and epidemiology as well as chemical, mechanical, electrical, industrial, civil, and environmental engineering. Because of differences in the formal education and training of these professionals, their ability to perform or review remediation design calculations varies considerably. For some, performing accurate design calculations for site remediation can become a seemingly insurmountable task.

Most, if not all, of the books dealing with site remediation provide only descriptive information on remedial technologies, and none, in my opinion, provides helpful guidance on illustrations of practical design calculations. This book covers important aspects of the major design calculations used in the field and also provides practical and relevant working information derived from the literature and my own hands-on experiences accumulated from consulting and teaching in this field. This book was written to address the current needs of practicing engineers, scientists, and legal experts who are employed by industry, consulting companies, law firms, and regulatory agencies in the field of soil and groundwater remediation. This book can also serve as a textbook or a reference book for undergraduate and graduate students who are pursuing a career in site remediation.

It has been 15 years since the release of the first edition in 1999. I appreciate (and enjoy) the feedback that I have received from many parts of the world. After being pushed many times, I finally have this second edition done. I sincerely hope this book becomes a useful tool for people working in site remediation. Your comments and suggestions are always welcome, and my email address is jkuo@fullerton.edu.

About the Author

Jeff (Jih-Fen) Kuo, PhD, worked in the environmental engineering industry for more than 10 years before joining the Department of Civil and Environmental Engineering at California State University, Fullerton in 1995. He gained his industrial experience while working at Groundwater Technology, Inc.; Dames and Moore; James M. Montgomery Consulting Engineers; Nanya Plastics; and the Sanitation Districts of Los Angeles County. His industrial experience in environmental engineering includes design and installation of air strippers, activated-carbon adsorbers, soil vapor extraction systems, bioremediation systems, and flare/catalytic incinerators for groundwater and soil remediation. He is also experienced in site assessment, including fate and transport analysis of toxic compounds in the environment; RI/FS work for landfills and Superfund sites; design of flanged connections to meet stringent fugitive emission requirements; development of emissions factors for air emissions from wastewater treatment; and conductance of application research on various wastewater treatment processes. Areas of research in environmental engineering include dechlorination of halogenated aromatics by ultrasound, fines/bacteria migration through porous media, biodegradability of heavy hydrocarbons, surface properties of composite mineral oxides, kinetics of activated-carbon adsorption, wastewater filtration, THM formation potential of ion-exchange resins, UV disinfection, sequential chlorination, nitrification/denitrification, removal of target compounds using nanoparticles, persulfate oxidation of persistent chemicals, microwave oxidation for wastewater treatment, landfill gas recovery and utilization, control technologies for greenhouse gases and fugitive methane emissions, and treatment of storm-water runoff.

Dr. Kuo earned a BS degree in chemical engineering from National Taiwan University, an MS in chemical engineering from the University of Wyoming, and an MS in petroleum engineering, and an MS and a PhD in environmental engineering from the University of Southern California. He is a professional civil, mechanical, and chemical engineer registered in California.

1

Introduction

1.1 Background and Objectives

The hazardous waste management business has steadily increased since the mid-1970s as public concern led to a vast range of new and stringent environmental regulations. With regard to groundwater and soil, a substantial amount of time and expense has been devoted to studying impacted sites, with much of the effort dedicated to litigation to determine the financially responsible parties. However, the focus has switched in recent years from litigation and site assessment to remediation. Site restoration usually proceeds through several phases and requires a concerted, multidisciplinary effort. Thus, remediation professionals come from a variety of technical and educational backgrounds, including geology, hydrology, chemistry, microbiology, meteorology, toxicology and epidemiology, as well as chemical, mechanical, electrical, industrial, civil, and environmental engineering. Because of differences in the formal education and training of these professionals, their ability to perform or review remediation design calculations varies considerably. For some, performing accurate design calculations for site remediation can become a seemingly insurmountable task.

Groundwater and soil remediation is more complicated than the conventional water and wastewater treatment because characteristics of soil and subsurface geology/hydrogeology greatly affect the implementability and effectiveness of a given technology. The absence of uniformly trained specialists is exacerbated by the continuously evolving remediation technologies. While up-to-date design information is sporadically published in the literature, it is usually theoretical in nature, and illustrative applications are rarely given. Most, if not all, of the books dealing with site remediation provide only descriptive information on remedial technologies, and none, in this author's opinion, provide helpful guidance on illustrations of practical design calculations.

Selection of a proper remedial alternative is site-specific. One needs to know the applicability and limitations of each technology before a smart decision can be made. In addition to knowing how a remedial technology works, it is more important to know why it may not work for an impacted site.

Without the proper information and education, environmental professionals can exert themselves, needlessly reinventing the wheel, so to speak, and err in design calculations. This book covers important aspects of the major design calculations used in the field and also provides practical and relevant working information derived from the literature and the author's own experience. Realistic examples are used liberally to illustrate the application of the design calculations. Many examples were designed to assist the readers in building the right concepts and common sense. This book was written to address the current needs of practicing engineers, scientists, and legal experts who are employed by industry, consulting companies, law firms, and regulatory agencies in the field of soil and groundwater remediation. This book can also serve as a textbook or a reference book for undergraduate and graduate students who are pursuing a career in site remediation.

1.2 Organization of the Book

In addition to this introductory chapter, the book is divided into the following six chapters:

Chapter 2: Site Assessment and Remedial Investigation. This chapter illustrates engineering calculations needed during site assessment and remedial investigation. It begins with simple calculations for estimating the amount of impacted soil excavated and that left in the vadose zone and the size of the plume in the aquifer. This chapter also covers necessary calculations to determine partitioning of compounds of concern (COCs) in different phases (soil, moisture, void, and free product), which is critical for design and implementation of remedial systems.

Chapter 3: Plume Migration in Aquifer and Soil. This chapter illustrates how to estimate the speeds of groundwater movement and plume migration. The reader will also learn how to interpret the aquifer test data and estimate the age of a groundwater plume.

Chapter 4: Mass-Balance Concept and Reactor Design. This chapter first introduces the mass-balance concept, followed by reaction kinetics, as well as types, configuration, and sizing of reactors. The reader will learn how to determine reaction-rate constants, removal efficiency, optimal arrangement of reactors, required residence time, and reactor size for one's specific applications.

Chapter 5: Vadose Zone Soil Remediation. This chapter provides important design calculations for commonly used *in situ* and *ex situ* soil remediation technologies, such as soil vapor extraction, soil washing, bioremediation, *in situ* chemical oxidation, low-temperature thermal desorption, and thermal destruction. Taking soil vapor extraction as an example, the book will guide the readers through design calculations for radius of influence, well spacing, air flow rate, extracted COC concentrations, effect of temperature on vapor flow, cleanup time, and sizing of vacuum pumps.

Chapter 6: Groundwater Remediation. This chapter starts with design calculations for capture zone and optimal well spacing. The rest of the chapter focuses on design calculations for commonly used *in situ* and *ex situ* groundwater remediation technologies, including activated carbon adsorption, air stripping, *in situ* and *ex situ* bioremediation, air sparging, biosparging, chemical precipitation, *in situ* chemical oxidation, and advanced oxidation processes.

Chapter 7: VOC-Laden Air Treatment. Remediation of impacted soil and groundwater often results in transferring organic COCs into the air phase. Development and implementation of an air emission control strategy are an integral part of the overall remediation program. This chapter illustrates design calculations for commonly used off-gas treatment technologies, including activated carbon adsorption, direct incineration, catalytic incineration, internal combustion (IC) engines, and biofiltration.

1.3 How to Use This Book

This book is constructed to provide a comprehensive coverage of commonly used soil and groundwater remediation technologies. It is written in a cookbook style and user-friendly format. Both SI and US customary units are used throughout the book, and unit conversions are frequently given. Examples are given following the design equations. Some of the examples are provided to illustrate important design concepts. One of the best ways to use the book is to glance through the entire book first, by reading the text and skimming the problem statement and discussion, and revisit the specific topics in detail later when related design calculations are to be made.

2

Site Assessment and Remedial Investigation

2.1 Introduction

The initial step, often the most critical one, of a typical soil and/or ground-water remediation project is to define the problem. It is accomplished by site assessment and remedial investigation (RI).

Site assessment (also referred to as *site characterization*) is to understand what has happened at a site. When site remediation is deemed necessary, RI will be employed. RI activities consist of additional site characterization and data collection. The data are needed in making engineering decisions on control of plume migration and selection of remedial alternatives. The common questions to be answered by the RI activities include the following:

- What media (surface soil, vadose zone, underlying aquifer, air) have been impacted?
- Where is the plume located in each impacted medium?
- What are the vertical and areal extents of the plume?
- What are the concentration levels of compounds of concern (COCs)?
- How long has the plume been there?
- Where is the plume going?
- Has the plume gone beyond the property boundary?
- How fast will the plume go?
- What are the on-site sources of the COCs?
- Are there potential off-site sources to this plume (now and/or in the past)?

Subsurface contamination from spills and leaky underground storage tanks (USTs) creates environmental conditions that usually require corrective remedies. The COCs may be present in a combination of the following locations and phases:

Vadose zone
- Vapors in the void
- Free product in the void
- Dissolved in the soil moisture
- Adsorbed onto the soil grains
- Floating on top of the capillary fringe (for light nonaqueous-phase liquids [LNAPLs])

Underlying aquifer
- Dissolved in the groundwater
- Adsorbed onto the aquifer material
- Coexisting with groundwater in the pores as free product or sitting on top of the bedrock (for dense nonaqueous-phase liquids [DNAPLs])

Common RI activities may include:

1. Removal of source(s) of contamination, such as leaky USTs
2. Installation of soil borings
3. Installation of groundwater monitoring wells
4. Collection and analysis of soil samples
5. Collection and analysis of groundwater samples
6. Collection of groundwater elevation data
7. Conduction of aquifer testing
8. Removal of impacted soil that may serve as a contamination source to the aquifer

Through these activities, the following data may be collected:

1. Types of COCs present in the vadose zone and underlying aquifer
2. Concentrations of COCs in the collected soil and groundwater samples
3. Vertical and areal extents of the plumes in the vadose zone and underlying aquifer
4. Vertical and areal extents of the free product (LNAPLs and DNAPLs)
5. Soil characteristics including types, density, porosity, moisture content, etc.

6. Groundwater elevations

7. Drawdown data from aquifer tests

Using these collected data, engineering calculations are then performed to assist in site remediation. Common engineering calculations include:

1. Mass and volume of soil excavated during tank removal

2. Mass and volume of impacted soil left in the vadose zone

3. Mass of COCs left in the vadose zone

4. Mass and volume of the free product (LNAPLs and DNAPLs)

5. Size of the dissolved plume in the aquifer

6. Mass of COCs present in the aquifer (dissolved and adsorbed)

7. Hydraulic gradient and groundwater flow direction

8. Hydraulic conductivity of the aquifer

This chapter describes all the above-needed engineering calculations except for the last two, which will be covered in Chapter 3. Discussion will also be presented with regard to calculations related to site assessment activities, including cuttings from soil borings and purge water from groundwater sampling. The last part of this chapter describes the "partitioning" of COCs in different phases. A good understanding of the partitioning phenomenon of COCs is critical for evaluation of the fate and transport of COCs in subsurface and for selection of remedial alternatives.

2.2 Determination of Extent of Contamination

2.2.1 Mass and Concentration Relationship

As mentioned earlier, COCs may be present in different media (i.e., soil, water, or air) and in different phases (i.e., vapor, dissolved, adsorbed, or free product). In environmental engineering applications, people commonly express concentrations of COCs in parts per million (ppm), parts per billion (ppb), or parts per trillion (ppt).

Although these concentration units are commonly used, some people may not fully realize that "1 ppm," for example, does not mean the same for liquid, solid, and air samples. For the liquid and solid phases, the ppm unit is on a mass-per-mass basis. One ppm stands for one part mass of a compound per million parts mass of the media containing it. Soil containing 1 ppm benzene means that every gram of soil contains 1 microgram (μg) of benzene, i.e., 10^{-6} g benzene per gram of soil (1 μg/g), or 1 mg benzene per kilogram of soil (1 mg/kg).

For the liquid phase, 1 ppm benzene means that 1 μg of benzene is dissolved in 1 g of water, or 1 mg benzene per kilogram water. Since it is usually more convenient to measure the liquid volume than its mass, and 1 kg of water has a volume of approximately 1 L under ambient conditions, people commonly use "1 ppm" for "1 mg/L compound concentration in liquid." It should be noted that 1 ppm (1 mg/kg or 1 mg/L) = 1,000 ppb (1,000 μg/kg or 1,000 μg/L) = 1,000,000 ppt (1,000,000 nanogram/kg or 1,000,000 ng/L). In addition, a concentration of 1% by weight (10,000 mg/kg) would be 10,000 ppm, since 1/100 = 10,000/1,000,000.

For the vapor phase, the story is totally different. The unit of ppm is on a volume-per-volume basis (or a mole-per-mole basis per se). The unit of ppm by volume (ppmV) is commonly used. One ppmV of benzene in the air means one part volume of benzene in 1 million parts volume of air space. To convert ppmV into mass concentration units, which is often needed in remediation work, we can use the following formula:

$$1\ \text{ppmV} = \frac{MW}{22.4}\ [\text{mg/m}^3]\quad \text{at } T = 0°\text{C}$$

$$= \frac{MW}{24.05}\ [\text{mg/m}^3]\quad \text{at } T = 20°\text{C} \qquad (2.1)$$

$$= \frac{MW}{24.5}\ [\text{mg/m}^3]\quad \text{at } T = 25°\text{C}$$

or

$$1\ \text{ppmV} = \frac{MW}{359} \times 10^{-6}\ [\text{lb/ft}^3]\quad \text{at } T = 32°\text{F}$$

$$= \frac{MW}{385} \times 10^{-6}\ [\text{lb/ft}^3]\quad \text{at } T = 68°\text{F} \qquad (2.2)$$

$$= \frac{MW}{392} \times 10^{-6}\ [\text{lb/ft}^3]\quad \text{at } T = 77°\text{F}$$

where MW is the molecular weight of the compound, and the number in the denominator of each equation above is the molar volume of an ideal gas at that temperature and one atmosphere. For example, the volume of an ideal gas is 22.4 L per gram-mole at 0°C, or 359 ft³ per pound-mole at 32°F and $P = 1$ atm.

Let us determine the conversion factors between ppmV and mg/m³ or lb/ft³, using benzene (C_6H_6) as an example ($P = 1$ atm). The MW of benzene is 78.1; therefore, 1 ppmV of benzene is the same as:

$$1\,\text{ppmV benzene} = \frac{78.1}{24.05} = 3.25 \text{ mg/m}^3 \quad \text{at } 20°C$$

$$= \frac{78.1}{24.5} = 3.19 \text{ mg/m}^3 \quad \text{at } 25°C$$

$$= \frac{78.1}{385} = 0.203 \times 10^{-6} \text{ lb/ft}^3 \quad \text{at } 68°F \ (20°C)$$

$$= \frac{78.1}{392} = 0.199 \times 10^{-6} \text{ lb/ft}^3 \quad \text{at } 77°F \ (25°C)$$

(2.3)

For comparison, the mass concentration of 1 ppmV perchloroethylene (PCE, C_2Cl_4, MW = 165.8) at 20°C and $P = 1$ atm is equal to:

$$1\,\text{ppmV PCE} = \frac{165.8}{24.05} = 6.89 \text{ mg/m}^3 \quad \text{at } 20°C$$

$$= \frac{165.8}{385} = 0.431 \times 10^{-6} \text{ lb/ft}^3 \quad \text{at } 68°F \ (20°C)$$

(2.4)

From this practice, we learn that the conversion factors are different among compounds because of the differences in molecular weight. The mass concentration of 1 ppmV PCE is twice as large as that of 1 ppmV benzene (6.89 vs. 3.25 mg/m³ at 20°C). In addition, the conversion factor for a compound is temperature dependent because its molar volume varies with temperature. The higher the temperature is, the smaller the mass concentration would be for the same ppmV value.

In remediation design, it is often necessary to determine the mass of a COC present in a medium. It can be found from the COC concentration and the amount of the medium containing the COC. The procedure for such calculations is simple, but slightly different for liquid, soil, and air phases. The differences mainly come from the units of concentration.

Let us start with the simplest case in which a liquid is impacted with a dissolved COC. Dissolved COC concentration in the liquid (C) is typically expressed in mass of COC/volume of liquid, such as mg/L; therefore, mass of the COC dissolved in liquid can be obtained by multiplying the concentration with the volume of liquid (V_l):

Mass of COC in liquid = (liquid volume) × (liquid concentration) = $(V_l)(C)$
(2.5)

When a soil sample is sent to a laboratory for analysis, it usually contains some moisture. The sample would be weighed first. The measured value includes the weight of the dry soil and that of the associated moisture. If the soil is impacted, the COCs should be present on the surface of the soil grains (as adsorbed) as well as in the soil moisture (as dissolved). Both the adsorbed and dissolved COCs in this soil sample would then be extracted and quantified as a whole. The COC concentration in soil (X) would be reported in the unit of "mass of COC/mass of soil," such as milligrams per kilogram. The mass of a COC in soil can be obtained by multiplying its concentration in soil with the mass of soil (M_s):

$$\text{Mass of COC in soil} = (\text{mass of soil}) \times (\text{COC concentration in soil}) = (M_s)(X)$$
(2.6)

The mass of soil can be estimated as the multiplication product of the volume and bulk density of the soil. Bulk density is the mass of a material divided by the total volume it occupies. For civil engineering practices, the reported values of bulk density are often on a dry-soil basis, i.e., "dry" bulk density (= mass of dry soil ÷ volume as a whole). However, the COC concentration in soil is usually based on the mass of wet soil (soil + moisture). Consequently, the bulk density used to calculate the soil mass for subsequent estimation of COC mass should be the "wet" bulk density (= mass of dry soil plus moisture ÷ volume as a whole). The "wet" bulk density is also referred to as "total" bulk density. In this book, ρ_t is the symbol for total bulk density and ρ_b is the symbol for (dry) bulk density. The mass of COC in soil can also be found as:

$$\text{Mass of COC in soil} = [(\text{soil volume})(\text{total bulk density})]$$
$$\times (\text{COC concentration in soil})$$
$$= [(V_s)(\rho_t)](X) = (M_s)(X)$$
(2.7)

COC concentration in air (G) is often expressed in vol/vol (such as ppmV) or in mass/vol (such as mg/m^3). In calculation of mass, we need to convert the concentration into the mass/vol basis using Equation (2.1) or (2.2). Mass of the COC in air can then be obtained by multiplying the mass concentration with the volume of the air (V_a):

$$\text{Mass of COC in air} = (\text{air volume}) \times (\text{COC concentration in mass/volume})$$
$$= (V_a)(G)$$
(2.8)

Example 2.1: Mass and Concentration Relationship

Which of the following media contains the largest amount of xylenes [$C_6H_4(CH_3)_2$]?

(a) 1 million gallons of water containing 10 ppm of xylene
(b) 100 cubic yards of soil (total bulk density = 1.8 g/cm^3) having 10 ppm of xylenes

(c) An empty warehouse ($200' \times 50' \times 20'$) containing 10 ppmV xylenes in air ($T = 20°C$; see Equation 2.1)

Solution:

(a) Mass of xylenes in liquid = (liquid volume)(liquid concentration)

$= [(1,000,000 \text{ gal})(3.785 \text{ L/gal})](10 \text{ mg/L})$

$= (3.785 \times 10^6)(10) = 3.79 \times 10^7 \text{ mg}$

(b) Mass of xylenes in soil = [(soil volume)(total bulk density)](COC concentration in soil)

$= \{[(100 \text{ yd}^3)(27 \text{ ft}^3/\text{yd}^3)(30.48 \text{ cm/ft})^3] \times [(1.8 \text{ g/cm}^3)(\text{kg}/1000\text{g})]\}$ (10 mg/kg)

$= (1.37 \times 10^5)(10) = 1.37 \times 10^6 \text{ mg}$

(c) Mass of xylenes in air:

MW of xylenes $[C_6H_4(CH_3)_2] = (12)(6) + (1)(4) + [12 + (1)(3)](2) = 106$ g/mole

At $T = 20°C$ and $P = 1$ atm

1 ppmV of xylenes = (MW of xylene/24.05) mg/m^3

$= (106/24.05) \text{ mg/m}^3 = 4.407 \text{ mg/m}^3$

10 ppmV of xylenes = (10 ppmV)[4.407 (mg/m^3)/ppmV]

$= (10)(4.407) = 44.07 \text{ mg/m}^3$

Mass of xylenes in air = (air volume)(vapor concentration)

$= [(200 \times 50 \times 20 \text{ ft}^3)(0.3048 \text{ m/ft})^3](44.07 \text{ mg/m}^3)$

$= (5.66 \times 10^3)(44.07) = 2.5 \times 10^5 \text{ mg}$

$\therefore$ The liquid contains the largest amount of xylene.

Discussion:

1. To convert a ppmV value to a mass concentration basis, we need to specify the corresponding temperature and pressure.
2. It should be noted that xylenes have three isomers, i.e., *ortho-*, *meta-*, and *para-*xylenes (*o-*, *m-*, *p-*xylenes).

Example 2.2: Mass and Concentration Relationship (in SI units)

Which of the following media contains the largest amount of toluene $[C_6H_5(CH_3)]$?

(a) 5,000 m³ of water containing 5 ppm of toluene
(b) 5,000 m³ of soil (total bulk density = 1,800 kg/m³) having 5 ppm of toluene
(c) An empty warehouse (indoor space = 5,000 m³) with 5 ppmV toluene in air ($T = 25°C$; see Equation 2.1)

Solution:

(a) Mass of COC in liquid = (liquid volume)(dissolved concentration)
$$= [(5,000 \text{ m}^3)(1,000 \text{ L/m}^3)](5 \text{ mg/L})$$
$$= (5 \times 10^6)(5) = 2.5 \times 10^7 \text{ mg}$$

(b) Mass of COC in soil = [(soil volume)(total bulk density)](COC concentration in soil)
$$= [(5,000 \text{ m}^3)(1,800 \text{ kg/m}^3)](5 \text{ mg/kg})$$
$$= (9.0 \times 10^6)(5) = 4.5 \times 10^7 \text{ mg}$$

(c) Mass of COC in air:
MW of toluene $[C_6H_5(CH_3)] = (12)(7) + (1)(8) = 92$ g/mole
At $T = 25°C$ and $P = 1$ atm,
$$5 \text{ ppmV of toluene} = (5 \text{ ppmV})[(\text{MW of toluene}/24.5)$$
$$((\text{mg/m}^3)/\text{ppmV})]$$
$$= (5)(92/24.5) = 18.76 \text{ mg/m}^3$$

Mass of COC in air = (air volume)(vapor concentration)
$$= [5,000 \text{ m}^3](18.76 \text{ mg/m}^3) = 9.38 \times 10^4 \text{ mg}$$

∴   The soil contains the largest amount of toluene.

Discussion:

1. Using SI units appears to be easier in these types of calculations. However, engineers, at least in the United States, need to master unit conversions in their job assignments because US customary units are still commonly used in the workplace.

2. With the same volume of 5,000 m³ and the same concentration of 5 ppm, the amounts in these three media are quite different.

3. Please be aware that the equations for ppmV to mass concentration conversion are different between this example and Example 2.1 because the temperatures are different (25°C vs. 20°C).

Example 2.3: Mass and Concentration Relationship

If an adult drinks water containing 10 ppb benzene and inhales air containing 10 ppbV benzene a day, which system (ingestion or inhalation) is exposed to more benzene? Note that the typical water intake rate is 2.0 L/day and the air inhalation rate is 15.2 m³/day for an adult [13].

Solution:

(a) Benzene ingested daily = (2 L/day)(10 × 10⁻³ mg/L) = 0.02 mg/day

(b) Benzene inhaled daily:

MW of benzene (C_6H_6) = (12)(6) + (1)(6) = 78 g/mole

At $T = 20°C$ and $P = 1$ atm,

10 ppbV benzene = (10 × 10⁻³)(78/24.05) mg/m³ = 0.0324 mg/m³

Benzene inhaled daily = (15.2 m³/day)(0.0324 mg/m³) = 0.49 mg/day

∴ The inhalation system is exposed to more benzene.

Discussion:

1. 1 ppb = 0.001 ppm

2. The pressure was not specified, and $P = 1$ atm was used (a reasonable assumption).

Example 2.4: Mass and Concentration Relationship

A glass bottle containing 900 mL of methylene chloride $(CH_2Cl_2$, specific gravity = 1.335) was accidentally left uncapped over a weekend in a poorly ventilated room (5 m × 6 m × 3.6 m). On the following Monday, it was found that two-thirds of methylene chloride had volatilized from the bottle.

For the worst-case scenario (i.e., all the volatilized methylene chloride stayed in the room, with no air exchange with the outside), would the concentration in the air exceed the Occupational Safety and Health Administration (OSHA's) eight-hour time-weighted average (TWA) permissible exposure limit (PEL) of 25 ppmV and the short-term exposure limit (STEL) of 125 ppmV?

Solution:

(a) Mass of methylene chloride volatilized = (liquid volume) × (density)

= [(2/3)(900 mL)](1.335 g/mL)

= (600)(1.335) = 801 g = 8.01 × 10⁵ mg

(b) Vapor concentration in mass/vol = (mass) ÷ (volume)

= (8.01 × 10⁵ mg) ÷ [(5 m)(6 m)(3.6 m)]

= (8.01 × 10⁵) ÷ (108) = 7,417 mg/m³

(c)   Vapor concentration in ppmV:
        MW of methylene chloride $[CH_2Cl_2] = (12) + (1)(2) + (35.5)(2)$
$$= 85 \text{ g/mole}$$

At $T = 20°C$ and $P = 1$ atm,
        1 ppmV of methylene chloride $= (85/24.05) \text{ mg/m}^3$
$$= 3.53 \text{ mg/m}^3$$

Vapor concentration in vol/vol $= 7,417 \text{ mg/m}^3 \div [3.53 \ (\text{mg/m}^3)/$
$$\text{ppmV}] = 2,100 \text{ ppmV}$$

∴   It would exceed the PEL and the STEL.

Discussion:

1. Specific gravity is the ratio of the density of a substance to the density of a reference substance, most commonly water at $4°C$ ($= 1 \text{ g/cm}^3 = 9.807 \text{ kN/m}^3$).
2. STEL is based on a sampling period of 15 min.
3. This calculation illustrates that a relatively small amount of COC emission could yield unhealthy air quality.

Example 2.5: Mass and Concentration Relationship

A boy went into a site and played with dirt impacted with ethyl benzene $(C_6H_5C_2H_5)$. During his stay at the site, he inhaled 2 m³ of air containing 10 ppbV of ethyl benzene and ingested a mouthful (≈ 1 cm³) of soil containing 5 ppm ethyl benzene. Which system (ingestion or inhalation) is exposed to more ethyl benzene? Assume the total bulk density of soil is equal to 1.8 g/cm³.

Solution:

(a)   Ethyl benzene inhaled
        MW of ethyl benzene $(C_6H_5C_2H_5) = (12)(8) + (1)(10) = 106$ g/mole
        At $T = 20°C$ and $P = 1$ atm,
        10 ppbV of ethyl benzene $= (10 \times 10^{-3})(106/24.05) \text{ mg/m}^3$
$$= 0.0441 \text{ mg/m}^3$$
        Mass of ethyl benzene inhaled
        $=$ (air volume)(vapor concentration)
        $= (2 \text{ m}^3)(0.0441 \text{ mg/m}^3) = 0.0882 \text{ mg}$

(b)   Ethyl benzene ingested
        $=$ [(soil volume)(total bulk density)](COC concentration in soil)
        $= [(1 \text{ cm}^3)(1.8 \text{ g/cm}^3)(1 \text{ kg}/1,000 \text{ g})](5 \text{ mg/kg})$
        $= (1.8 \times 10^{-3})(5) = 0.0090 \text{ mg}$

∴   The inhalation system is exposed to more ethyl benzene.

Discussion:

1. The default air intake rate for children of 6–8 years old is 10 m³/ day (ATSDR, 2005). With a total air intake of 2 m³, this child would have been playing out there for a few hours.

2. The average soil ingestion rate for children is 200 mg/day, while that for pica children is 5,000 mg/day [13]. (This rate should only be used when assessing acute exposure situations.)

3. Benzene/toluene/ethyl benzene/xylenes (B/T/E/X) are the main COCs in gasoline because of their toxicity. Please note that the chemical formulas of ethyl benzene and xylenes are the same (and so are their molecular weights).

Example 2.6: Gas Concentration in ppmV

The vapor pressure of mercury at $T = 25°C$ and $P = 1$ atm is 0.0017 mm-Hg. If mercury is allowed to evaporate to equilibrium in an enclosed space, determine the theoretical mercury concentration (in ppm) in air.

Solution:

(a) Mole fraction of mercury in air $= P_{mercury}/P_{total}$

$= (0.0017 \text{ mm-Hg})/(760 \text{ mm-Hg}) = 2.24 \times 10^{-6}$

(b) Vapor concentration of mercury $=$ mole fraction of mercury in air

$= 2.24 \times 10^{-6} = 2.24$ parts per million $= 2.24$ ppm (or ppmV)

Discussion:

1. The calculation for this question is relatively simple. To correctly answer the question, we need to have the right concept about vapor concentrations in the units of ppm (or ppmV).

2. An engineer should be familiar with units commonly used for pressure. One atmosphere $= 1.013 \times 10^5$ Pa $= 101.3$ kPa $= 1.013$ bar $= 1,013$ mbar $= 760$ mm-Hg $= 760$ torr $= 29.92$ in.-Hg $= 14.696$ lb/in.² (psi) $= 33.9$ ft-$H_2O = 10.33$ m-H_2O.

Example 2.7: Conversion of Gas Concentrations between ppmV and Mass Concentration

The National Ambient Air Quality Standard (NAAQS) for nitrogen dioxide (NO_2) is 100 ppb (1-h average). A dispersion modeling analysis of NO_2 emissions from a source shows a maximum ambient receptor concentration of 180 μg/m³. The receptor elevation is 6,000 ft; the barometric pressure is

24.0 in.-Hg; and the ambient temperature is 68°F. Determine the 1-hr average NAAQS value for NO_2 at that location (in $\mu g/m^3$).

Solution:

(a) At $T = 68°F$ and $P = 24.0$ in.-Hg, molar volume of an ideal gas

$$= (22.4 \text{ L/gmole})(29.92/24.0)[(460 + 68)/(460 + 32)]$$

$$= (22.4)(1.25)(1.07) \text{ L/gmole} = 29.97 \text{ L/gmole}$$

(b) Molecular weight of $NO_2 = (14)(1) + (16)(2) = 46$ g/mole

Under this ambient condition, 0.100 ppmV of $NO_2 = (0.100)(MW$ of $NO_2/29.97)$ mg/m^3

$$= (0.100)(46/29.97)$$

$$= 0.153 \text{ mg/m}^3 = 153 \ \mu g/m^3 \ (< 180 \ \mu g/m^3)$$

∴   The maximum ambient receptor concentration exceeds the 1-hr average NAAQS for NO_2.

Discussion:

1. One may encounter questions of this nature in professional engineers exams. In this example, both values of pressure and temperature are included in the conversion between ppmV and mass concentration, while $P = 1$ atm was assumed in the previous examples.

2. To calculate the molar volume of an ideal gas, we can always use the Ideal Gas Law ($PV = nRT$) with a proper value of the ideal, or universal, gas constant (R), see Table 2.1 for values of the universal gas constant in different units. The approach here started with 22.4 L/gmole, which is the molar volume of an ideal gas at $T = 0°C$ and $P = 1$ atm. (This value is a good one for us to memorize.) Since the volume is proportional to temperature and inversely proportional to pressure, the relationship, $V_2/V_1 = (T_2/T_1)(P_1/P_2)$, is valid.

3. Temperature used, in Ideal Gas Law–related calculations, should be the absolute temperature in degrees Kelvin (K) or degrees Rankine (°R). Note: T (in K) $= T$ (in °C) $+ 273.15$, and T (in °R) $= T$ (in °F) $+ 459.67$. Also, T (in °R) $= 1.8 \times T$ (in K).

TABLE 2.1

Values of the Universal Gas Constant (R)

$R = 82.05$ (cm^3·atm)/(g mol)(K)	$= 83.14$ (cm^3·bar)/(g mol)(K)
$R = 8.314$ (J)/(g mol)(K)	$= 1.987$ (cal)/(g mol)(K)
$R = 0.7302$ (ft^3·atm)/(lb mol)(R)	$= 10.73$ (ft^3·psia)/(lb mol)(R)
$R = 1{,}545$ (ft·lb$_f$)/(lb mol)(R)	$= 1.986$ (Btu)/(lb mol)(R)

Example 2.8: Bulk Densities, Water Content, and Degree of Water Saturation

The average specific gravity of soil grains at a site is 2.65; porosity is equal to 0.40; and water content (weight of moisture/weight of dry soil) is 0.12. Determine the (dry) bulk density, total bulk density, volumetric water content, and degree of water saturation of the soil.

Solution:

(a) Basis: soil volume = 1 m³ of soil

Total pore volume of the soil = (soil volume) × (porosity of soil)
$$= (1)(0.4) = 0.4 \text{ m}^3$$

Volume occupied by the soil grains
$$= (\text{soil volume}) - (\text{its pore volume})$$
$$= 1 - 0.4 = 0.6 \text{ m}^3$$

Mass of the dry soil = (volume occupied by the soil grains)
$$\times (\text{density of the soil grains})$$
$$= (0.6 \text{ m}^3)(2,650 \text{ kg/m}^3) = 1,590 \text{ kg}$$

Bulk density (dry) = (mass of the dry soil) ÷ (soil volume)
$$= 1,590 \text{ kg} \div 1 \text{ m}^3$$
$$= 1,590 \text{ kg/m}^3 = 1.59 \text{ g/cm}^3 \ (= 99.2 \text{ lb/ft}^3)$$

(b) Mass of water/moisture in soil
$$= (\text{water content}) \times (\text{mass of the dry soil})$$
$$= (0.12)[(1,590 \text{ kg/m}^3)(1 \text{ m}^3)] = 190.8 \text{ kg}$$

Mass of the wet soil
$$= (\text{mass of water}) + (\text{mass of the dry soil})$$
$$= 190.8 + 1,590 = 1,781 \text{ kg}$$

Total bulk density = (mass of the wet soil) ÷ (soil volume)
$$= 1,781 \text{ kg} \div 1 \text{ m}^3 = 1,781 \text{ kg/m}^3$$
$$= 1.78 \text{ g/cm}^3 \ (= 111.1 \text{ lb/ft}^3)$$

(c) Volume of water = (mass of water) ÷ (density of water)
$$= (190.8 \text{ kg}) \div (1,000 \text{ kg/m}^3) = 0.19 \text{ m}^3$$

Volumetric water content = (volume of water) ÷ (soil volume)
$$= (0.19 \text{ m}^3) \div 1.0 \text{ m}^3 = 19\%$$

(d) Degree of water saturation
$$= (\text{volume of water}) \div (\text{total pore volume})$$
$$= (0.19 \text{ m}^3) \div (0.4 \text{ m}^3) = 47.5\%$$

Discussion:

1. One m³ of soil was used as the "basis" in this example. Other volumes (e.g., 1 ft³) can also be used, and the results should be the same.

2. Although *mass* and *weight* are different, these two terms are often used interchangeably in this book (and in many other engineering articles).

3. Many equations that relate these parameters can be found in technical articles. However, the procedure used in this example, without using any of those equations, was to develop a better understanding of the concepts and definitions.

4. As expected, the value of the total bulk density (1.78 g/cm³ or 111.1 lb/ft³) is larger than that of the (dry) bulk density (1.59 g/cm³ or 99.2 lb/ft³).

5. In civil engineering practices, water content is usually on a gravimetric basis. However, in environmental engineering applications, volumetric water content and degree of water saturation are more commonly used. In this example, for a water content of 0.12, water occupies 19% of the total soil volume and 47.5% of the pore volume (air occupies the balance, i.e., 52.5% of the pore volume).

2.2.2 Amount of Soil from Tank Removal or from Excavation of the Impacted Area

Removal of USTs typically involves soil excavation. If the excavated soil is clean (i.e., free of COCs or below the permissible levels), it may be reused as backfill materials or disposed of in a sanitary landfill. On the other hand, if it is impacted, it needs to be treated or disposed of in a hazardous waste landfill. For either case, a good estimate of soil volume and/or mass is necessary. If feasible, we should separate the apparently impacted soil from the clean soil by putting them into separate piles to save the subsequent treatment/disposal costs. Using a portable instrument, such as a photo-ionization detector (PID), flame-ionization detector (FID), or organic vapor analyzer (OVA), would help us to make the decision.

The excavated soil is typically stored on site first in stockpiles. The amount of excavated soil from tank removal can be determined from measurement of the volumes of the stockpiles. However, the shapes of these piles are irregular, and this makes the measurement and subsequent calculations difficult. An easier and more accurate alternative is

Step 1: Measure the dimensions of the tank pit.

Step 2: Calculate the volume of the tank pit from the measured dimensions.

Step 3: Determine the number and volumes of the USTs removed.

Step 4: Subtract the total volume of the USTs from the volume of the tank pit.

Step 5: Multiply the value from Step 4 with a soil fluffy factor.

Information needed for this type of calculation

- Dimensions of the tank pit (from field measurements)
- Number and volumes of the USTs removed (from drawings or field observation)
- Bulk density of soil (from measurement or estimate)
- Soil fluffy factor (from estimate)

Example 2.9: Determine the Mass and Volume of Soil Excavated from a Tank Pit

Two 5,000-gallon USTs and one 6,000-gallon UST were removed. The excavation resulted in a tank pit of $50' \times 24' \times 18'$. The excavated soil was stockpiled on site. The total bulk density of soil *in situ* (before excavation) is 1.8 g/cm^3 and that of soil in the stockpiles is 1.5 g/cm^3. Estimate the volume and mass of the excavated soil.

Solution:

Volume of the tank pit = $(50')(24')(18') = 21{,}600$ ft^3

Total volume of the USTs = $(2)(5{,}000) + (1)(6{,}000) = 16{,}000$ gal

$$= (16{,}000 \text{ gal})(\text{ft}^3/7.48 \text{ gal}) = 2{,}139 \text{ ft}^3$$

Volume of soil in the tank pit before removal
= (volume of tank pit) – (volume of USTs)
= $21{,}600 - 2{,}139 = 19{,}461$ ft^3

Volume of soil excavated (in the stockpiles)
= (volume of soil in the tank pit) × (fluffy factor)
= $(19{,}461)(1.2) = 23{,}353$ ft^3
= $(23{,}353 \text{ ft}^3)(\text{yd}^3/27 \text{ ft}^3) = 865$ yd^3

Mass of soil excavated = (volume of the soil in the tank pit)(total bulk density of soil *in situ*)

= (volume of the soil in the stockpile)(total bulk density of soil in the stockpile)

Total bulk density of soil *in situ* = (1.8 g/cm³)[(62.4 lb/ft³)/(1 g/cm³)]
$$= (1.8)(62.4) = 112.32 \text{ lb/ft}^3$$

Total bulk density of soil in stockpiles = (1.5)(62.4) = 93.6 lb/ft³
Mass of soil excavated
$$= (19{,}461 \text{ ft}^3)(112.32 \text{ lb/ft}^3) = 2{,}185{,}800 \text{ lb}$$
$$= (23{,}353 \text{ ft}^3)(93.6 \text{ lb/ft}^3) = 2{,}185{,}800 \text{ lb} = 1{,}093 \text{ tons}$$

Discussion:

1. The fluffy factor of 1.2 is to take into account the loosening of soil after being excavated from subsurface (the *in situ* soil is usually more compacted). A fluffy factor of 1.2 means that the volume of soil increases 20% from *in situ* to the stockpiles. On the other hand, the bulk density of soil in the stockpiles would be smaller than that of *in situ* soil as the result of becoming "loose" after excavation.

2. The calculated mass of the excavated soil should be the same; regardless, the volume of soil in the tank pit or that in the stockpiles is used.

3. For the US customary system, one (short) ton = 2,000 lb, while in SI, one (long) ton = 1,000 kg (which is equivalent to 2,200 lb).

4. Sizes of USTs at gasoline stations nowadays are typically larger and in the neighborhood of 10,000 gallons.

Example 2.10: Mass and Concentration Relationship of Excavated Soil

A leaky 20 m³ underground storage tank was removed. The excavation resulted in a tank pit of 4 m × 4 m × 5 m (L×W×H), and the excavated soil was stockpiled on site. Three samples were taken from the pile, and the total petroleum hydrocarbon (TPH) concentrations were determined to be ND (not detectable, <100), 1,500, and 2,000 ppm. What is the amount of TPH in the pile? Express your answers in both kilograms and liters.

Solution:

Volume of the tank pit = (4)(4)(5) = 80 m³
Volume of soil in the tank pit before excavation
$$= \text{(volume of the tank pit)} - \text{(volume of the USTs)}$$
$$= 80 - 20 = 60 \text{ m}^3$$

Average TPH concentration = (100 + 1,500 + 2,000)/3 = 1,200 ppm
$$= 1{,}200 \text{ mg/kg}$$
Mass of TPH in soil = [(60 m³)(1,800 kg/m³)](1,200 mg/kg)
$$= (1.08 \times 10^5)(1{,}200) \text{ mg} = 1.30 \times 10^8 \text{ mg} = 130 \text{ kg}$$

Volume of TPH in soil = (mass of TPH) ÷ (density of TPH)

$$= (130 \text{ kg}) \div (0.8 \text{ kg/L}) = 162.5 \text{ L} = 41.9 \text{ gal}$$

Discussion:

1. The total bulk density of soil was assumed to be 1,800 kg/m³ (i.e., 1.8 g/cm³), and the density of TPH was assumed to be 0.8 kg/L (i.e., 0.8 g/cm³).

2. The TPH concentration for one of the three samples is below the detection limit of 100 ppm. Four approaches are commonly taken to deal with values below the detection limit: (1) use the detection limit as the value, (2) use half of the detection limit, (3) use zero, and (4) select a value based on a statistical approach (especially when multiple samples are taken and a few of them are below the detection limit). In this solution, a conservative approach was taken by using the detection limit as the concentration.

3. We may often find that some values in tables of technical articles are shown as ND. It would be better to show the corresponding detection limits of these samples, e.g., ND (<100 ppm), as in this example.

Example 2.11: Mass and Concentration Relationship of Excavated Soil

A leaky 1,000-gallon underground storage tank was removed. The excavation resulted in a tank pit of 12′ × 12′ × 15′ (L×W×H), and the excavated soil was stockpiled on site. Five samples were taken from the pile and analyzed for TPH using EPA method 8015.

Based on the laboratory results, an engineer at CSUF Consulting Company estimated that there were approximately 50 gallons of gasoline present in the stockpile. One of the five TPH values in the report was illegible, and the others were ND (<100), 1,000, 2,000, and 3,000 ppm, respectively. What is the missing value?

Solution:

Let x be the missing TPH value.

Average TPH concentration = (x + 100 + 1,000 + 2,000 + 3,000)/

5 (in mg/kg)

Mass of the impacted soil = [(12)(12)(15) − (1,000/7.48) ft³](112 lb/ft³)

$$= (2,026)(112) = 227,000 \text{ lb} = 103,000 \text{ kg}$$

Mass of TPH in soil = (volume of gasoline) × (density of gasoline)

$$= [(50 \text{ gal})(\text{ft}^3/7.48 \text{ gal})] \times [(50 \text{ lb/ft}^3)(\text{kg}/2.2 \text{ lb})]$$

$$= (6.68)(22.73) = 151.9 \text{ kg}$$

151.9 kg = (average TPH concentration) × (mass of the impacted soil)

$$= [(x + 6,100)/5 \text{ mg/kg}] \times [(103,000 \text{ kg})(\text{kg}/10^6 \text{ mg})]$$

$$x = \text{the unknown TPH concentration} = 1,264 \text{ mg/kg}$$

Discussion:

1. The total bulk density of soil was assumed to be 112 lb/ft^3, which is equal to 1.79 g/cm^3 (= 112/62.4)
2. The density of TPH was assumed to be 50 lb/ft^3, which is equal to 0.80 g/cm^3 (= 50/62.4)

2.2.3 Amount of Impacted Soil in the Vadose Zone

Chemicals that leaked from USTs might move beyond the tank pit. If subsurface contamination is suspected, soil borings are drilled to assess the extent of contamination in the vadose zone. Soil-boring samples are then taken at a fixed interval, e.g., every 5 or 10 ft, and analyzed for soil properties. Selected samples are submitted to certified laboratories and analyzed for COCs. From these data, a fence diagram is often developed to delineate the extent of the COC plume.

When selecting remedial alternatives, an engineer needs to know the vertical and areal extents of the plume, types of subsurface soil, types of COCs, mass and/or volume of the impacted soil, and mass of COCs in different phases. If the location of the plume is shallow (not deep from the ground surface level [gsl]) and the amount of the impacted soil is not extensive, excavation coupled with on-site aboveground treatment or off-site treatment/disposal may be a viable option. On the other hand, *in situ* remediation alternatives, such as soil venting, would be more favorable if the volume of the impacted soil is large and deep. Therefore, a good estimate of the amount of impacted soil left in the vadose zone is important for remediation considerations. This section describes the methodology for such calculations.

As mentioned, a fence diagram is often drawn to illustrate the vertical and areal extents of the plume. Based on the information from the diagram, the following procedure can be used to estimate the amount of the impacted soil in the vadose zone:

Step 1: Determine the area of the plume at each sampling depth, A_i.

Step 2: Determine the thickness interval for each area calculated above, h_i.

Step 3: Determine the volume of the impacted soil, V_s, using the following formula:

$$V_s = \sum_i A_i h_i \qquad (2.9)$$

Step 4: Determine the mass of the impacted soil, M_s, by multiplying V_s with the total bulk density of soil, ρ_t, as:

$$M_s = \rho_t \times V_s \qquad (2.10)$$

Information needed for this calculation

- The areal and vertical extents of the plume, h_i and A_i
- Total bulk density of the soil (*in situ*), ρ_t

To determine the mass and volume of the impacted water contained in a groundwater plume, the following procedure should be followed:

Step 1: Use Equation (2.9) to determine the size of the plume.

Step 2: Multiply the size from Step 1 by the aquifer porosity to obtain the volume of the impacted groundwater.

Step 3: Multiply the volume from Step 2 by groundwater density to obtain the mass of the impacted water.

Example 2.12: Determine the Amount of Impacted Soil in the Vadose Zone

For the project described in Example 2.9, after the USTs were removed, five soil borings were installed. Soil samples were taken every 5 ft below ground surface (bgs). The area of the plume at each soil sampling interval was determined as follows:

Depth (ft bgs)	Area of the Plume at that Depth (ft²)
15	0
20	350
25	420
30	560
35	810
40	0

Determine the volume and mass of the impacted soil left in the vadose zone.

Strategy:

The soil samples were taken and analyzed every 5 ft; therefore, each plume area represents the same depth interval. The sample taken at 20-ft depth represents the 5-ft interval from 17.5 to 22.5 ft (the mid-depth of the first two consecutive intervals to the mid-depth of the next two consecutive intervals). Similarly, the sample at 25-ft depth represents the 5-ft interval from 22.5 to 27.5 ft, and so on.

Solution:

Thickness intervals for all areas of the plume are the same at 5 ft.

Volume of the impacted soil (using Equation [2.9])

$$= (5 \text{ ft})(350 \text{ ft}^2) + (5\text{ft})(420 \text{ ft}^2) + (5 \text{ ft})(560 \text{ ft}^2) + (5 \text{ ft})(810 \text{ ft}^2)$$
$$= (1{,}750 + 2{,}100 + 2{,}800 + 4{,}050) \text{ ft}^3 = 10{,}700 \text{ ft}^3 = 396 \text{ yd}^3$$
$$\text{or} = (22.5 - 17.5)(350) + (27.5 - 22.5)(420) + (32.5 - 27.5)(560)$$
$$+ (37.5 - 32.5)(810) = 10{,}700 \text{ ft}^3$$

Assuming the total bulk density of soil is 112 lb/ft³, the mass of the impacted soil = (10,700 ft³)(112 lb/ft³)

$$= 1{,}198{,}400 \text{ lb} = 599 \text{ tons}$$

Example 2.13: Determine the Amount of Impacted Soil in the Vadose Zone

For the project described in Example 2.9, after the USTs were removed, five soil borings were installed. Soil samples were taken every 5 ft below ground surface (bgs). However, not all the samples were analyzed due to budget con-straints. The areas of the plume at a few depths were determined as follows:

Depth (ft bgs)	Area of the Plume at that Depth (ft²)
15	0
20	350
25	420
35	810
40	0

Determine the volume and mass of the impacted soil left in the vadose zone.

Strategy:

The depth intervals given are not all the same; therefore, each plume area represents a different depth interval. For example, the sample taken at 25-ft depth represents a 7.5-ft interval, from 22.5 ft to 30 ft.

Solution:

Volume of the impacted soil (using Equation [2.9])

$$= (5)(350) + (7.5)(420) + (7.5)(810) \text{ ft}^3$$
$$= 10{,}915 \text{ ft}^3 = 406 \text{ yd}^3$$
$$\text{or} = (22.5 - 17.5)(350) + (30 - 22.5)(420) + (37.5 - 30)(810) = 10{,}915 \text{ ft}^3$$

Assuming the total bulk density of soil is 112 lb/ft³, the mass of the impacted soil = (10,975 ft³)(112 lb/ft³)

$$= 1{,}229{,}200 \text{ lb} = 615 \text{ tons}$$

Example 2.14: Determine the Amount of Impacted Soil in the Vadose Zone (in SI units)

After the leaky USTs were removed, five soil borings were installed. Soil samples were taken every 2 m below ground surface (bgs). However, not all the samples were analyzed due to budget constraints. The areas of the plume at a few depths were determined as follows:

Depth (m bgs)	Area of the Plume at that Depth (m²)
6	0
8	35
10	42
14	81
16	0

Determine the volume and mass of the impacted soil left in the vadose zone.

Strategy:

The depth intervals given are not the same; therefore, each plume area represents a different depth interval. For example, the sample taken at 10-m depth represents a 3-m interval, from 9 to 12 m.

Solution:

Volume of the impacted soil (using Equation 2.9)

$$= (2)(35) + (3)(42) + (3)(81) \text{ m}^3$$
$$= 439 \text{ m}^3$$

$$\text{or} = (9-7)(35) + (12-9)(42) + (15-12)(81) = 439 \text{ m}^3$$

Assuming the total bulk density of soil is 1,800 kg/m³, the mass of the impacted soil = (439 m³)(1,800 kg/m³)

$$= 790{,}200 \text{ kg} = 790 \text{ tons}$$

2.2.4 Mass Fraction and Mole Fraction of Components in Gasoline

Gasoline is a common COC found in subsurfaces, usually the result of leaky USTs. Gasoline itself is a mixture of various hydrocarbons, and it may contain more than 200 different compounds. Some of them are lighter and more volatile than the others (lighter ends vs. heavier ends). Gasoline in soil samples is usually measured by EPA method 8015 as total petroleum hydrocarbon (TPH), using gas chromatography (GC). Diesel fuel is often measured by "modified" EPA method 8015 that takes into account the abundance of heavier ends in diesel fuel as compared to gasoline. Some gasoline

TABLE 2.2

Some Physicochemical Properties of B/T/E/X

	Formula	MW	Water Solubility (mg/L)	Vapor Pressure (mm-Hg)
Benzene	C_6H_6	78	1,780 @ 25°C	95 @ 25°C
Toluene	$C_6H_5(CH_3)$	92	515 @ 20°C	22 @ 20°C
Ethyl benzene	$C_6H_5(C_2H_5)$	106	152 @ 20°C	7 @ 20°C
Xylenes	$C_6H_4(CH_3)_2$	106	198 @ 20°C	10 @ 20°C

Source: [4]

constituents are more toxic than others. Benzene, toluene, ethyl benzene, and xylenes (B/T/E/X) are gasoline constituents of concern because of their toxicity. (Benzene is a known human carcinogen.) B/T/E/X compounds are often measured by EPA method 8020 or 8260.

To cut down the air pollution, many oil companies have developed so-called reformulated gasoline in which the benzene content is reduced [1]. Beginning in 2011, benzene content in all US gasoline was reduced to ≤0.62% (by volume) to comply with the Mobile Sources Air Toxics Rule [2, 3]. Some of the important physicochemical properties of B/T/E/X are tabulated in Table 2.2. (Note: Physical and chemical properties of chemical compounds can now be readily found from searching the Internet.) The Material Safety and Data Sheet (MSDS), now called Safety Data Sheet (SDS), is a good source for these types of data.

Sometimes, it is necessary to determine the composition, such as mass and mole fractions of important constituents, of the gasoline for the following reasons:

1. *Identification of potential responsible parties*: At a busy intersection having two or more gasoline stations, the free-floating product found beneath a site may not come from its USTs. Each brand of gasoline usually has its own distinct formula. Most oil companies have the capability to identify biomarkers in the gasoline or to determine if the composition of free-floating products matches their formula.

2. *Determination of health risk*: As mentioned, some gasoline constituents are more toxic than others, and these should be considered differently in a health-risk assessment.

3. *Estimation on the age of plume*: Some compounds are more volatile than others. The fraction of volatile constituents in a recent gasoline spill should be larger than that in an aged spill.

To determine mass fractions of components in gasoline, the following procedure can be used:

Step 1: Determine the mass of the mixture (i.e., TPH) and mass of each COC.

Step 2: Determine the mass fraction by dividing the mass of each COC with the mass of TPH.

To determine the mole fractions of components in gasoline, the following procedure can be used:

Step 1: Determine the mass of TPH and mass of each COC in the impacted soil.

Step 2: Determine the molecular weight of each COC.

Step 3: Determine the molecular weight of gasoline from the composition and molecular weights of all constituents. This procedure is tedious, and data may not be readily available. Assuming the molecular weight of gasoline to be 100, which is equivalent to that of heptane (C_7H_{16}), is relatively reasonable.

Step 4: Calculate the number of moles of each COC by dividing its mass with its molecular weight.

Step 5: Calculate the mole fraction of each COC by dividing the number of moles of each COC with the number of moles of TPH.

Information needed for this calculation

- Mass of the impacted soil
- Concentrations of the COCs
- Molecular weights of the COCs

Example 2.15: Mass and Mole Fractions of COCs in Gasoline

Three samples were taken from a soil pile (110 yd³) and analyzed for TPH (EPA method 8015) and for B/T/E/X (EPA method 8020). The average concentration of TPH is 1,000 mg/kg, and those of B/T/E/X are 20, 20, 20, and 20 mg/kg, respectively. Determine the mass and mole fractions of B/T/E/X in the gasoline. The total bulk density of the soil in the stockpile is 1.65 g/cm³.

Solution:

(a) Mass of the impacted soil

   = (volume of soil)(total bulk density)

$$= [(110 \text{ yd}^3)(27 \text{ ft}^3/\text{yd}^3)]\{(1.65 \text{ g/cm}^3)[62.4 \text{ lb/ft}^3/(1 \text{ g/cm}^3)]\}$$
$$= (2{,}970)(103) \text{ lb} = 305{,}900 \text{ lb} = 139{,}000 \text{ kg}$$

(b) Mass of COC in soil = (mass of soil)(COC concentration in soil)

Mass of TPH = $(139{,}000 \text{ kg})(1{,}000 \text{ mg/kg})$
$$= 1.39 \times 10^8 \text{ mg} = 1.39 \times 10^5 \text{ g}$$

Mass of benzene = $(139{,}000 \text{ kg})(20 \text{ mg/kg}) = 2.78 \times 10^6 \text{ mg}$
$$= 2.78 \times 10^3 \text{ g}$$

Mass of toluene = $(139{,}000 \text{ kg})(20 \text{ mg/kg}) = 2.78 \times 10^6 \text{ mg}$
$$= 2.78 \times 10^3 \text{ g}$$

Mass of ethyl benzene = $(139{,}000 \text{ kg})(20 \text{ mg/kg}) = 2.78 \times 10^6 \text{ mg}$
$$= 2.78 \times 10^3 \text{ g}$$

Mass of xylenes = $(139{,}000 \text{ kg})(20 \text{ mg/kg}) = 2.78 \times 10^6 \text{ mg}$
$$= 2.78 \times 10^3 \text{ g}$$

(c) Mass fraction of a COC = (mass of the COC) ÷ (mass of TPH)

Mass fraction of benzene = $(2.78 \times 10^3)/(1.39 \times 10^5) = 0.020 = 2.0\%$

Mass fraction of toluene = $(2.78 \times 10^3)/(1.39 \times 10^5) = 0.020 = 2.0\%$

Mass fraction of ethyl benzene = $(2.78 \times 10^3)/(1.39 \times 10^5) = 0.020$
$$= 2.0\%$$

Mass fraction of xylenes = $(2.78 \times 10^3)/(1.39 \times 10^5) = 0.020 = 2.0\%$

(d) Moles of a COC = (mass of the COC) ÷ (MW of the COC)

Moles of TPH = $(1.39 \times 10^5)/(100) = 1{,}390$ g-mole

Moles of benzene = $(2.78 \times 10^3)/(78) = 35.6$ g-mole

Moles of toluene = $(2.78 \times 10^3)/(92) = 30.2$ g-mole

Moles of ethyl benzene = $(2.78 \times 10^3)/(106) = 26.2$ g-mole

Moles of xylenes = $(2.78 \times 10^3)/(106) = 26.2$ g-mole

(e) Mole fraction of a COC
$$= (\text{moles of the COC})/(\text{moles of TPH})$$

Mole fraction of benzene = $(35.6)/(1{,}390) = 0.0256 = 2.6\%$

Mole fraction of toluene = $(30.2)/(1{,}390) = 0.0217 = 2.2\%$

Mole fraction of ethyl benzene = $(26.2)/(1{,}390) = 0.0189 = 1.9\%$

Mole fraction of xylenes = $(26.2)/(1{,}390) = 0.0189 = 1.9\%$

Discussion:

1. The mass fraction of each COC can also be directly determined from the ratio of the COC concentration and the TPH concentration. Using benzene as an example, mass fraction of benzene = (20 mg/kg)/(1,000 mg/kg) = 0.020 = 2.0%.

2. The mass fractions of B/T/E/X are all the same (2.0%) because they have the same concentration (20 mg/kg). On the other hand, their mole fractions are different because of the differences in their molecular weights.

2.2.5 Height of Capillary Fringe

The capillary fringe (or capillary zone) is a zone immediately above the water table of unconfined aquifers. It extends upward from the top of the water table due to the capillary rise of water. The capillary fringe often creates complications in site-remediation projects. In general, the size of the plume in the aquifer would be much larger than that in the vadose zone, because of the spread of the dissolved plume in the aquifer. If the water table fluctuates, the capillary fringe will move upward or downward with the water table. Consequently, the capillary fringe above the dissolved groundwater plume can become impacted. In addition, if free-floating product exists, the fluctuation of the water table will cause the free product to move vertically and laterally. The site remediation for this scenario will be more complicated and difficult. In addition, most of the commonly used technologies cannot effectively remediate the impacted capillary zones.

The height of capillary fringe at a site strongly depends on its subsurface geology. For pure water at 20°C in a clean glass tube, the height of capillary rise can be approximated by the following equation:

$$h_c = \frac{0.153}{r} \tag{2.11}$$

where h_c is the height of capillary rise in cm, and r is the radius of the capillary tube in cm. This formula can be used to estimate the height of the capillary fringe. As shown in Equation (2.11), the thickness of the capillary fringe will vary inversely with the pore size of the formation. Table 2.3 summarizes the information from two references with regard to capillary fringe. As the grain size becomes smaller, the pore radius often gets smaller, and the capillary rise increases. The thickness of the capillary fringe of a clayey aquifer can be as large as 10 ft.

TABLE 2.3

Typical Height of Capillary Fringe

Material	Grain Size (mm)[a]	Pore Radius (cm)[b]	Capillary Rise (cm)	
			Source A[a]	Source B[b]
Gravel				
Coarse	...	0.4	...	0.38
Fine	2–5	...	2.5	...
Sand				
Very coarse	1–2	...	6.5	...
Coarse	0.5–1	0.05	13.5	3.0
Medium	0.2–0.5	...	24.6	...
Fine	0.1–0.2	0.02	42.8	7.7
Silt				
	0.05–0.1	0.001	105.5	150
	0.02–0.05	...	200	...
Clay	...	0.0005	...	300

Source: [5, 6].

Example 2.16: Thickness of Capillary Fringe

A core sample was taken from an impacted unconfined aquifer and analyzed for pore size distributions. The effective pore radius was determined to be 5 μm. Estimate the thickness of the capillary fringe of this aquifer.

Solution:

Pore radius = 5×10^{-6} m = 5×10^{-4} cm

Using Equation (2.11), we obtain

Capillary rise = $(0.153)/(5 \times 10^{-4})$ = 306 cm = 3.06 m = 10.0 ft

Discussion:

1. Equation (2.11) is an empirical equation. The units for capillary rise and pore radius in this equation need to be in centimeters. By looking at the equation, having both units in centimeters does not seem to match. However, the constant (0.153) has taken care of the unit conversions. If other units are used, the value of the constant would be different.

2. The calculated value (306 cm) for the capillary rise is essentially the same as the value (300 cm) listed in Table 2.3 for clay with pore radius of 0.005 mm.

2.2.6 Estimating the Mass and Volume of the Free-Floating Product

The LNAPL product leaked from a UST may accumulate on the top of the capillary fringe of a water-table (unconfined) aquifer or on the top of the upper confining layer of a confined aquifer to form a free-product layer. For site remediation, it is often necessary to estimate the volume or mass of this free-floating product. The thickness of the free product found in the monitoring wells had been directly used to calculate the volume of free product outside the wells. However, these calculated values are seldom representative of the actual free-product volume existing in the formation.

It is now well known that the thickness of free product found in the formation (the actual thickness) is much smaller than that floating on top of the water in the monitoring well (the apparent thickness). Using the apparent thickness, without any adjustment, to estimate the volume of free product may lead to an overestimate of the free-product volume and overdesign of the remediation system. The overestimate of free product in the RI phase may cause difficulties in obtaining an approval for final site closure, because the remedial action can never recover the full amount of free product reported in the site assessment report.

Factors affecting the difference between the actual thickness and the apparent thickness include the densities (or specific gravity) of the free product and the groundwater as well as the characteristics of the formation (especially the pore sizes). Several approaches have been presented in the literature to correlate these two thicknesses. Ballestero, Fiedler, and Kinner [7] developed an equation using heterogeneous fluid flow mechanics and hydrostatics to determine the actual free-product thickness in an unconfined aquifer. The equation is

$$t_g = t(1 - SG) - h_a \tag{2.12}$$

where
- t_g = actual (formation) free-product thickness
- t = apparent (wellbore) free-product thickness
- SG = specific gravity of the free product
- h_a = distance from the bottom of the free product to the water table.

If no further data for h_a are available, average wetting capillary rise can be used as h_a. Information on capillary rise can be found in Section 2.2.5.

To estimate the actual thickness of free product, the following procedure can be used:

Step 1: Determine the specific gravity of free product. (The specific gravity of gasoline can be reasonably assumed as 0.75 to 0.80, if no additional information is available.)

Step 2: Measure the apparent thickness of the free product inside the well.

Step 3: Calculate the actual thickness of free product in the formation by inserting values of these parameters into Equation (2.12).

Information needed for this calculation

- Specific gravity (or density) of the free product, SG
- Measured thickness of the free product in the well, t
- Capillary rise, h_c

To determine the mass and volume of the free-floating product, the following procedure can be used:

Step 1: Determine the areal extent of the free-floating product.

Step 2: Estimate the true thickness of the free-floating product.

Step 3: Calculate the volume of the free-floating product by multiplying the area with the true thickness and the effective porosity of the formation.

Step 4: Calculate the mass of the free-floating product by multiplying the volume with its density.

Information needed for this calculation

- Areal extent of the free-floating product
- True thickness of the free-floating product
- Effective porosity of the formation
- Density (or specific gravity) of the free-floating product

Example 2.17: Determine the True Thickness of the Free-Floating Product

A recent survey of a groundwater monitoring well showed a 75-in.-thick layer of gasoline floating on top of the water. The density of gasoline is 0.8 g/cm³, and the thickness of the capillary fringe above the water table is 1 ft. Estimate the actual thickness of the free-floating product in the formation.

Solution:

Using Equation (2.12), we obtain:

Actual free-product thickness in the formation (t_g)

$$= (75)(1 - 0.8) - 12 = 3 \text{ in.}$$

Discussion:

1. Specific gravity is the ratio of the density of a substance to the density of a reference substance (commonly, water at 4°C).
2. As shown in this example, the actual thickness of the free product is only 3 in., while the apparent thickness within the monitoring well is much larger at 75 in. (a 25-fold difference).

Example 2.18: Estimate the Mass and Volume of the Free-Floating Product

Recent groundwater-monitoring results at an impacted site indicate that the areal extent of the free-floating product has an approximately rectangular shape of 50 ft × 40 ft. From the apparent thicknesses of free product in four monitoring wells inside the plume, the true thicknesses of free product in the vicinities of these four wells were estimated to be 2, 2.6, 2.8, and 3 ft, respectively. The effective porosity of the subsurface is 0.35. Estimate the mass and volume of the free-floating product present at the site. Assume the specific gravity of the free-floating product is equal to 0.8.

Solution:

(a) The areal extent of the free-floating product
$$= (50')(40') = 2,000 \text{ ft}^2$$

(b) The average thickness of the free-floating product
$$= (2 + 2.6 + 2.8 + 3)/4 = 2.6 \text{ ft}$$

(c) The volume of the free-floating product
$$= (\text{volume of the free-floating product zone}) \times (\text{effective porosity of the formation})$$
$$= [(\text{area})(\text{thickness})] \times (\text{effective porosity of the formation})$$
$$= [(2,000 \text{ ft}^2)(2.6 \text{ ft})](0.35)$$
$$= (5,200)(0.35) = 1,820 \text{ ft}^3 = 13,610 \text{ gal}$$

(d) Mass of the free-floating product
$$= (\text{volume of the free-floating product})(\text{density of the free-floating product})$$
$$= (1,820 \text{ ft}^3)\{0.8 \text{ g/cm}^3) \times [62.4 \text{ lb/ft}^3/(1 \text{ g/cm}^3)]\}$$
$$= 90,854 \text{ lb} = 41,300 \text{ kg}$$

Discussion:

Effective porosity should be used instead of *porosity* for these types of estimates. The *effective porosity* represents the portion of pore space that contributes to flow of the fluid (i.e., free product here) through the porous medium.

2.2.7 Determination of the Extent of Contamination: A Comprehensive Example

This subsection presents a comprehensive example of calculations related to common site-assessment activities.

Example 2.19: Determination of the Extent of Contamination: A Comprehensive Example

A gasoline station is located in the greater Los Angeles Basin within the floor plain of the Santa Ana River. The site is underlain primarily with coarser-grained river deposit alluvium. Three 5,000-gallon steel tanks were excavated and removed in May of 2013, with the intention that they would be replaced with three double-walled tanks within the same excavation.

During the tank removal, it was observed that the tank backfill soil exhibited a strong gasoline odor. Based on visual observations, the fuel hydrocarbon in the soil appeared to have been caused by overspillage during filling at unsealed fill boxes or minor piping leakage at the eastern end of the tanks. The excavation resulted in a pit of 20' × 30' × 18' (*L×W×H*). The excavated soil was stockpiled on site. Four samples were taken from the piles and analyzed for TPH using EPA method 8015. The TPH concentrations were ND (not detectable, <100), 200, 400, and 800 ppm.

The tank pit was then backfilled with clean dirt and compacted (the three new USTs were installed at a different location that was not impacted). Six vertical soil borings (two within the excavated area) were drilled to characterize the subsurface geological condition and to delineate the plume. The borings were drilled using the hollow-stem-auger method. Soil samples were taken by a 2-in.-diameter split-spoon sampler with brass soil sample retainers every 5 ft bgs. The water table is at 50 ft bgs, and all the borings were terminated at 70 ft bgs. All the borings were then converted to 4-in. groundwater monitoring wells.

Selected soil samples from the borings were analyzed for TPH (EPA method 8015) and B/T/E/X (EPA method 8020). The analytical results indicated that the samples from the borings outside the excavated area were all ND. The other results are listed here:

Boring no.	Depth (ft)	TPH (ppm)	Benzene (ppb)	Toluene (ppb)
B1	25	800	10,000	12,000
B1	35	2,000	25,000	35,000
B1	45	500	5,000	7,500
B2	25	ND (<10)	ND (<100)	ND (<100)
B2	35	1,200	10,000	12,000
B2	45	800	2,000	3,000

It was also found that free-floating gasoline product was present in the two monitoring wells located within the excavated area. The apparent thickness of the product in each of these two wells was converted to its actual thickness in the formation, and they are 1 and 2 ft, respectively. The effective porosity and total bulk density of soil are 0.35 and 1.8 g/cm^3, respectively.

Assuming that the leakage impacted a rectangular block of soil, defined by the bottom of the tank pit and the water table, with length and width equal to those of the tank pit, estimate the following:

(a) Total volume of the soil stockpiles (in cubic yards)
(b) Mass of TPH in the stockpiles (in kilograms)
(c) Volume of the impacted soil left in the vadose zone (in cubic meters)
(d) Mass of TPH, benzene, and toluene in the vadose zone (in kilograms)
(e) Mass fraction and mole fraction of benzene and toluene in the leaked gasoline
(f) Volume (in gallons) and mass (in kilograms) of the free product
(g) Total volume of gasoline leaked (in gallons) [Note: Neglect the dissolved phase in the underlying aquifer.]

Solution:

(a) Assuming a fluffy factor of 1.2, total volume of the soil stockpiles
   $$= \text{(volume of the soil } in \ situ)\text{(soil fluffy factor)}$$
   $$= [\text{(volume of tank pit)} - \text{(volume of USTs)}]\text{(soil fluffy factor)}$$
   $$= [(20' \times 30' \times 18') - (3)(5{,}000 \text{ gal})(\text{ft}^3/7.48 \text{ gal})](1.2)$$
   $$= (8{,}795 \text{ ft}^3)(1.2) = 10{,}550 \text{ ft}^3 = 391 \text{ yd}^3$$

(b) Mass of TPH in the stockpiles
   $$= [(V)(\rho_t)](X) = (M_s)(X)$$
   where
   $$X = (10 + 200 + 400 + 800)/4 = 352.5 \text{ mg/kg}$$
   $$\rho_t \ (in \ situ \ \text{soil}) = (1.8 \text{ g/cm}^3) \ (28{,}317 \text{ cm}^3/\text{ft}^3)(\text{kg}/1{,}000 \text{ g})$$
   $$= 51.0 \text{ kg/ft}^3 = 1{,}376 \text{ kg/yd}^3$$
   $$\rho_t \ \text{(stockpiles)} = \rho_t \ (in \ situ \ \text{soil}) \div \text{(soil fluffy factor)}$$
   $$= 51.0 \text{ kg/ft}^3 \div 1.2 = 42.5 \text{ kg/ft}^3$$
   ∴ Mass of TPH in the stockpiles
   $$= [(8{,}795 \text{ ft}^3)(51.0 \text{ kg/ft}^3)](352.5 \text{ mg/kg})$$
   $$= [(10{,}550 \text{ ft}^3)(42.5 \text{ kg/ft}^3)](352.5 \text{ mg/kg})$$
   $$= 1.58 \times 10^8 \text{ mg} = 158 \text{ kg}$$

(c) Volume of impacted soil left in the vadose zone
    = (area)(thickness)
    = $(20' \times 30')(50' - 18')$
    = $(600)(32) = 19{,}200$ ft^3 = 544 m^3 (Note: 1 ft^3 = 0.0283 m^3)

(d) Mass of TPH, benzene, and toluene in the vadose zone
    = $(V)(\rho_t)(X) = (M_s)(X)$ or using a more precise approach:

$$\sum_i (A_i)(h_i)(\rho_t)(X_i)$$

	Average Concentration (mg/kg)	Mass (kg)
TPH	$(800 + 2000 + 500 + 10 + 1200 + 800)/6 = 885$	$(19{,}200)(51)(885)/1{,}000{,}000 = 866$
Benzene	$(10 + 25 + 5 + 0.1 + 10 + 2)/6 = 8.68$	$(19{,}200)(51)(8.68)/1{,}000{,}000 = 8.50$
Toluene	$(12 + 35 + 7.5 + 0.1 + 12 + 3)/6 = 11.6$	$(19{,}200)(51)(11.6)/1{,}000{,}000 = 11.34$

(e) Mass fraction and mole fraction of benzene and toluene:

	Mass (kg)	Mass Fraction	MW	kg-mole	Mole Fraction
TPH	866	...	100	$866/100 = 8.66$	...
Benzene	8.50	$8.50/866 = 0.0098$	78	$8.50/78 = 0.109$	$0.109/8.66 = 0.0126$
Toluene	11.34	$11.3/866 = 0.0109$	92	$11.3/92 = 0.123$	$0.123/8.66 = 0.0142$

(f) Volume of free-floating product
    = $(h)(A)(\phi)$
    = $[(1 + 2)/2][(20 \times 30)](0.35) = 315$ ft$^3 \times$ (7.48 gal/ft^3) = 2,360 gal
    Mass of free-floating product
    = $(V)(\rho) = [(2{,}360$ gal$)(3.785$ L/gal$)](0.75$ kg/L$) = 6{,}700$ kg

(g) Total volume of gasoline leaked
    = Sum of those in the excavated soil, vadose zone, free product, and dissolved phase
    = $158 + 866 + 6{,}700 = 7{,}724$ kg (neglecting the dissolved phase)
    = $7{,}724$ kg $\div$ (0.75 kg/L) = 10,300 L = 2,720 gal

Discussion:

Estimation of COC mass in impacted aquifers is covered in Section 2.4.

2.3 Soil Borings and Groundwater Monitoring Wells

This section deals with calculations related to installation of soil borings and groundwater-monitoring wells and purging before groundwater sampling.

2.3.1 Amount of Cuttings from Soil Boring

The cuttings from soil borings are often temporarily stored on site in 55-gallon drums before final disposal. It becomes necessary to estimate the amount of cuttings and the number of drums needed. The calculation is relatively straightforward and easy, as shown here.

To estimate the amount of cuttings from soil boring, the following procedure can be used:

Step 1: Determine the diameter of the boring, d_b.

Step 2: Determine the depth of the boring, h.

Step 3: Calculate the volume of the cuttings using the following formula:

$$\text{Volume of cuttings} = \sum \left(\frac{\pi}{4} d_b^2 \right)(h)(\text{fluffy factor}) \qquad (2.13)$$

Information needed for this calculation

- Diameter of each boring, d_b
- Depth of each boring, h
- Soil fluffy factor

Example 2.20: Amount of Cuttings from Soil Boring

Four 10-in. boreholes are to be drilled to 50 ft below ground surface level for installation of 4-in. groundwater monitoring wells. Estimate the amount of soil cuttings and determine the number of 55-gallon drums needed to store the cuttings.

Solution:

(a) Volume of cuttings from each boring

$= [(\pi/4)(10/12)^2](50)(1.2) = 32.7 \text{ ft}^3$

Volume of cuttings from all four borings

$= (4)(32.7) = 131 \text{ ft}^3$

(b) Number of 55-gallon drums needed

$= (131 \text{ ft}^3)(7.48 \text{ gal/ft}^3) \div (55 \text{ gal/drum}) = 17.8 \text{ drums}$

Answer: Eighteen 55-gallon drums needed.

2.3.2 Amount of Packing Material and/or Bentonite Seal

Packing and seal materials need to be purchased and shipped to the site before installation of monitoring wells. A good estimate of the amount of packing material and bentonite seal is necessary for site assessment.

To estimate the packing and seal materials needed, the following procedure can be used:

Step 1: Determine the diameter of the boring, d_b.

Step 2: Determine the diameter of the well casing, d_c.

Step 3: Determine the thickness interval of the well packing or bentonite seal, h

Step 4: Calculate the volume of the packing or bentonite seal using the following formula:

$$\text{Volume of packing or bentonite needed} = \frac{\pi}{4}\left(d_b^2 - d_c^2\right)h \qquad (2.14)$$

Step 5: Determine the mass of the well packing or bentonite needed by multiplying its volume with its total bulk density.

Information needed for this calculation

- Diameter of the borehole, d_b
- Diameter of the casing, d_c
- Thickness interval of the packing or the bentonite seal, h
- Total bulk density of the packing or bentonite seal, ρ_t

Example 2.21: Amount of Packing Materials Needed

The four monitoring wells in Example 2.20 are to be installed 15 ft into the groundwater aquifer. The wells are to be perforated (0.02-in. slot opening) 15 ft below and 10 ft above the water table. Monterey Sand #3 is selected as the packing material. Estimate the number of 50-lb sand bags needed for this application. Assume the total bulk density of sand is equal to 1.8 g/cm³ (112 lb/ft³).

Solution:

(a) Packing interval for each well

= perforation interval + 1 ft = (10 + 15) + 1 = 26 ft

Volume of sand needed for each well

= {(π/4)[(10/12)² – (4/12)²]}(26) = 11.9 ft³

Volume of sand needed for four wells = (4)(11.9) = 47.6 ft³

(b) Number of 50-lb sand bags needed

$$= (47.6 \text{ ft}^3)(112 \text{ lb/ft}^3) \div (50 \text{ lb/bag}) = 107 \text{ bags}$$

Answer: 107 bags are needed.

Discussion:

1. Packing interval should be slightly larger than the perforation interval.
2. We should add an additional 10% to the estimate of sand usage as a safety factor to take into consideration that borehole shape would not be a perfect cylinder.

Example 2.22: Amount of Bentonite Seal Needed

The four monitoring wells in Example 2.21 are to be sealed with 5 ft of bentonite below the top grout. Estimate the number of 50-lb bags of bentonite needed for this application. Assume the total bulk density of bentonite is equal to 1.8 g/cm^3 (112 lb/ft^3).

Solution:

(a) Volume of bentonite needed for each well

$$= \{(\pi/4)[(10/12)^2 - (4/12)^2]\}(5) = 2.29 \text{ ft}^3$$

Volume of bentonite needed for four wells

$$= (2.29)(4) = 9.16 \text{ ft}^3$$

(b) Number of 50-lb bentonite bags needed

$$= (9.16 \text{ ft}^3)(112 \text{ lb/ft}^3) \div (50 \text{ lb/bag}) = 20.5 \text{ bags}$$

Answer: 21 bags are needed.

Discussion:

We should add an additional 10% to the estimate of bentonite usage as a safety factor to take into consideration that borehole shape would not be a perfect cylinder.

2.3.3 Well Volume for Groundwater Sampling

Purging is the process of removing stagnant water from a monitoring well before sampling groundwater. The stagnant volume includes the water inside the well casing and in the sand/gravel packing. A few parameters are often monitored, such as conductivity, pH, and temperature, to ensure that they reach a consistent endpoint before sampling. The purge volume is site specific and depends heavily on the subsurface geology. A rule of thumb is that purging three to five well volumes before groundwater sampling can be a starting point. The purged water is often impacted and needs to be treated, stored, and disposed of offsite. A good estimate of the volume of purged water is necessary for site assessment.

To estimate the amount of purged water, the following procedure can be used:

Step 1: Determine the diameter of the boring, d_b.

Step 2: Determine the diameter of the well casing, d_c.

Step 2: Determine the depth of the water in the well, h.

Step 3: Calculate the well volume using the following formula:

Well volume = volume of the groundwater enclosed inside the well casing
+ volume of the groundwater in the pore space of the packing

$$\text{Well volume} = \left[\frac{\pi}{4} d_c^2 \right] h + \left[\frac{\pi}{4} \left(d_b^2 - d_c^2 \right) h \right] \phi \qquad (2.15)$$

Information needed for this calculation:

- Diameter of the borehole, d_b
- Diameter of the casing, d_c
- Effective porosity of the packing, ϕ
- Depth of the well water, h

Example 2.23: Well Volume for Groundwater Sampling

The water depth inside each of the four monitoring wells in Example 2.21 was measured to be 14.5 ft. Three well volumes need to be purged out before sampling. Calculate the amount of purge water and also the number of 55-gallon drums needed to store the water. Assume the effective porosity of the well packing is equal to 0.40.

Solution:

(a) Well volume
    $= [(\pi/4) \times (4/12)^2 \times (14.5)] + \{(\pi/4) \times [(10/12)^2 - (4/12)^2]$
    $\times (14.5)\} \times (0.4)$
    $= 3.92 \text{ ft}^3$

(b) Three well volumes
    $= (3)(3.92) = 11.8 \text{ ft}^3 = 88 \text{ gal for each well}$

(c) Number of 55-gal drums needed for each well
   = [(11.8 ft³)(7.48 gal/ft³)] ÷ (55 gal/drum) = 1.6 drums
(d) Total number of 55-gal drums needed
   = (1.6)(4) = 6.4 drums

Answer: Seven 55-gallon drums are needed.

2.4 Mass of COCs Present in Different Phases

Once an NAPL enters a vadose zone, it may end up in four different phases. COCs may leave the free product and enter the void space. The COCs in the void and in the free product, in contact with the soil moisture, may get dissolved or absorbed in the liquid. Those COCs that enter the soil moisture may adsorb onto the soil grains. In other words, the NAPL can partition into four phases: (1) free product, (2) vapor in the void, (3) dissolved constituent in the soil moisture, and (4) adsorbed onto the soil grains. The concentrations of the COCs in the air void, in the soil moisture, and on the soil grains are interrelated and are affected greatly by the presence or absence of the free product. The partition of the COCs in these four phases has a great impact on the fate and transport of the COCs and the required site-remediation effort. A good understanding of this partition phenomenon is necessary to implement cost-effective alternatives for the site cleanup.

In this section, we first discuss the vapor concentration resulting from the presence of free product in the pores (Section 2.4.1). We then describe the relationship between the COC concentration in the liquid and that in the air (Section 2.4.2). The relationship between the COC concentration in the liquid and that on the soil grains is covered in Section 2.4.3. The relationship among the liquid, vapor, and solid concentrations is discussed in Section 2.4.4. The procedure to determine the partition of COC in these phases is discussed in Section 2.4.5.

2.4.1 Equilibrium between Free Product and Vapor

When a liquid is in contact with air, molecules in the liquid will tend to enter the air space as vapor, via volatilization. The vapor pressure of a liquid is the pressure exerted by its vapor at equilibrium. It is often reported in millimeters of mercury. (Note: 760 mm-Hg = 760 torr = 1 atm = 1.013×10^5 N/m² = 1.013×10^5 Pascal = 1.013 bar = 14.696 psi.) and varies greatly with temperature. In general, the higher the temperature is, the higher the vapor pressure will be. Several equations have been established to correlate the vapor pressure and temperature; the Clausius-Clapeyron equation is commonly used. This equation assumes that the enthalpy of vaporization is independent of temperature and is expressed as

$$\ln\frac{P_1^{sat}}{P_2^{sat}} = -\frac{\Delta H^{vap}}{R}\left[\frac{1}{T_1} - \frac{1}{T_2}\right] \tag{2.16}$$

where P^{sat} is the vapor pressure of the compound as a pure liquid, T is the absolute temperature, R is the universal gas constant, and ΔH^{vap} is the enthalpy of vaporization, which can be found in chemistry handbooks and references such as Lide [8]. Table 2.1 lists the values of the universal gas constant in various units.

The Antoine equation, which describes the relationship between vapor pressure and temperature, is widely used and has the following form:

$$\ln P^{sat} = A - \frac{B}{T+C} \tag{2.17}$$

where A, B, and C are the Antoine constants, which can be found in chemistry handbooks such as Reid, Prausnitz, and Poling [9].

For an ideal solution, the vapor–liquid equilibrium follows Raoult's law as:

$$P_A = (P^{vap})(x_A) \tag{2.18}$$

where
P_A = partial pressure of compound A in the vapor phase
P^{vap} = vapor pressure of compound A as a pure liquid
x_A = mole fraction of compound A in the solution

The partial pressure is the pressure that a compound would exert if all other gases were not present. This is equivalent to the mole fraction of the compound in the gas phase multiplied by the entire pressure of the gas. Raoult's law holds only for ideal solutions. In the dilute aqueous solutions commonly found in environmental applications, Henry's law, which will be discussed in the next section, is more suitable.

Example 2.24: Vapor Concentration in Void with Presence of Free Product

Benzene leaked from a UST at a site and entered the vadose zone. Estimate the maximum benzene concentration (in ppmV) in the pore space of the sub-surface. The temperature of the subsurface is 25°C.

Solution:

From Table 2.2, the vapor pressure of benzene is 95 mm-Hg at 25°C.

95 mm-Hg = (95 mm-Hg) ÷ (760 mm-Hg/1 atm) = 0.125 atm

The partial pressure of benzene in the pore space is 0.125 atm (125,000 × 10^{-6} atm), which is equivalent to 125,000 ppmV.

Discussion:

This 125,000-ppmV value is the vapor concentration in equilibrium with the pure benzene solution. The equilibrium can occur in a confined space or a stagnant phase. If the system is not totally confined, the vapor tends to move away from the source and creates a concentration gradient (i.e., the vapor concentration decreases with the distance from the liquid). However, in the vicinity of the solution, the vapor concentration would be at or near this equilibrium value.

Example 2.25: Using the Clausius-Clapeyron Equation to Estimate the Vapor Pressure

The enthalpy of vaporization of benzene is 33.83 kJ/mol [8] and the vapor pressure of benzene at 25°C is 95 mm-Hg (from Table 2.2). Estimate the vapor pressure of benzene at 20°C using the Clausius-Clapeyron equation.

Solution:

Heat of vaporization = 33.83 kJ/mol = 33,830 J/mol
R = 8.314 (J)/(g mol)(K) from Table 2.1
Using Equation (2.17), we obtain

$$\ln \frac{95}{P_2^{sat}} = -\frac{33,830}{8.314}\left[\frac{1}{(273+25)} - \frac{1}{(273+20)}\right]$$

P^{sat} of benzene at 20°C = 75 mm-Hg

Discussion:

As expected, the vapor pressure of benzene at 20°C is smaller than that at 25°C. The difference is approximately 20% (75 vs. 95 mm-Hg).

Example 2.26: Using the Antoine Equation to Estimate the Vapor Pressure

The empirical constants of the Antoine equation for benzene are [9]

A = 15.9008
B = 2788.51
C = −52.36

Estimate the vapor pressure of benzene at 20°C and at 25°C using the Antoine equation.

Solution:

(a) Use Equation (2.17), at 20°C

$$\ln P^{sat} = A - \frac{B}{T+C} = 15.9008 - \frac{2788.51}{(293-52.36)} = 4.322$$

So, $P^{vap} = 75.3$ mm-Hg

(b) Use Equation (2.17), at 25°C

$$\ln P^{sat} = A - \frac{B}{T+C} = 15.9008 - \frac{2788.51}{(298-52.36)} = 4.557$$

So, $P^{vap} = 95.3$ mm-Hg

Discussion:

1. The calculated benzene vapor pressure, 95.3 mm-Hg (at 25°C), is essentially the same as that in Table 2.2, 95 mm-Hg.
2. The calculated benzene vapor pressure, 75.3 mm-Hg (at 20°C), is essentially the same as that in Example 2.25, 75 mm-Hg, which uses the Clausius-Clapeyron equation.

Example 2.27: Vapor Concentration in the Void with Presence of Free Product

An industrial solvent, consisting of 50% (by weight) toluene and 50% ethyl benzene, leaked from a UST at a site and entered the vadose zone. Estimate the maximum toluene and ethyl benzene concentrations (in ppmV) in the void of the subsurface. The temperature of the subsurface is 20°C.

Solution:

From Table 2.2, the vapor pressure of toluene (C_7H_8, MW = 92) is 22 mm-Hg and that of ethyl benzene (C_8H_{10}, MW = 106) is 7 mm-Hg at 20°C.

For 50% by weight of toluene, the corresponding percentage by mole (i.e., mole fraction) will be

= (moles of toluene) ÷ [(moles of toluene)

+ (moles of ethyl benzene)] × 100

= (50/92) ÷ [(50/92) + (50/106)] × 100

= 53.5%

The partial pressure of toluene in the void can be estimated from Equation (2.18)

$$= (P^{vap})(x_A)$$

$$= (22)(0.535) = 11.78 \text{ mm-Hg}$$
$$= 0.0155 \text{ atm} = 15{,}500 \text{ ppmV}$$

The partial pressure of ethyl benzene in the pore space can also be estimated from Equation (2.18)

$$= (7)(1 - 0.535) = 3.25 \text{ mm-Hg}$$
$$= 0.0043 \text{ atm} = 4{,}300 \text{ ppmV}$$

Discussion:

The vapor concentrations are those in equilibrium with the solvent. The equilibrium can occur in a confined space or a stagnant phase. If the system is not totally confined, the vapor tends to move away from the source and creates a concentration gradient (the vapor concentration decreases with the distance from the solvent). However, in the vicinity of the solvent, the vapor concentration would be at or near the equilibrium value.

2.4.2 Liquid–Vapor Equilibrium

The compound in the void space of the vadose zone may enter the soil moisture via dissolution or absorption. Equilibrium conditions exist when the rate of the compound entering the soil moisture equals the rate of compound volatilizing from the soil moisture.

Henry's Coefficient. Henry's law is used to describe the equilibrium relationship between the liquid concentration and the vapor concentration. At equilibrium, the partial pressure of a gas above a liquid is proportional to the concentration of the chemical in the liquid. Henry's law can be expressed as

$$P_A = H_A\, C_A \tag{2.19}$$

where

P_A = partial pressure of compound A in the vapor phase
H_A = Henry's constant of compound A
C_A = concentration of compound A in the liquid phase

This equation shows a linear relationship between the liquid and vapor concentrations. The higher the liquid concentration is, the higher the vapor concentration will be. It should be noted that in some air pollution books or references, Henry's law is written as $C_A = H_A P_A$. This Henry's constant is the inverse of the one used in this book and most of the site-remediation articles.

Henry's law can also be expressed in the following form:

$$G = HC \tag{2.20}$$

where C is the COC concentration in the liquid phase and G is the corresponding concentration in the gas phase.

TABLE 2.4

Conversion Table for Henry's Constant

Desired Unit for Henry's Constant	Conversion Equation
atm/M, or atm·L/mole	$H = H'RT$
atm·m³/mole	$H = H'RT/1{,}000$
M/atm	$H = 1/(H'RT)$
atm/(mole fraction in liquid), or atm	$H = (H'RT)[1{,}000\gamma/W]$
(mole fraction in vapor)/(mole fraction in liquid)	$H = (H'RT)[1{,}000\gamma/W]/P$

Source: [10].

Note: H' = Henry's constant in the dimensionless form
  γ = specific gravity of the solution (1 for dilute solution)
  W = equivalent molecular weight of solution (18 for dilute aqueous solution)
  R = 0.082 atm/(K)(M)
  T = system temperature in Kelvin
  P = system pressure in atm (usually = 1 atm)
  M = solution molarity in (g·mol/L)

Henry's law has been widely used in various disciplines to describe the distribution of solute in the vapor phase and the liquid phase. The units of the Henry's constant (or Henry's law constant) reported in the literature vary considerably. The units commonly encountered include atm/mole fraction, atm/M, M/atm, atm/(mg/L), and dimensionless. When inserting the value of Henry's constant into Equations (2.19) and (2.20), it is important to check if its units match those of the other two parameters. Engineers with whom I have conferred normally use the units they are familiar with and often have difficulties in performing the necessary unit conversions. For your convenience, Table 2.4 is the conversion table for Henry's constant. Use of Henry's constant in dimensionless form has increased significantly. It should be noted that it is not a "(mole fraction)/(mole fraction)" dimensionless unit. The actual meaning of the Henry's constant in a dimensionless format is (concentration in the vapor phase)/(concentration in the liquid phase), which can be [(mg/L)/(mg/L)]. To be more precise, it has a unit of "(unit volume of liquid)/(unit volume of air)."

Henry's constant of any given compound varies with temperature. The Henry's constant is practically the ratio of the vapor pressure and the solubility, provided that both are measured at the same temperature, that is

$$H = \frac{\text{Vapor pressure}}{\text{Solubility}} \tag{2.21}$$

This equation implies that the higher the vapor pressure, the larger the Henry's constant is. In addition, the lower the solubility (or less soluble compound), the larger Henry's constant will be. For most organic compounds, the vapor pressure increases and the solubility decreases with temperature. Consequently, Henry's constant, as defined by Equations (2.20) or (2.21), should increase with temperature.

TABLE 2.5

Physicochemical Properties of Common COCs

Compound	MW (g/mole)	H (atm/M)	P^{vap} (mm-Hg)	D (cm²/s)	Log K_{ow}	Solubility (mg/L)	T (°C)
Benzene	78.1	5.55	95.2	0.092	2.13	1,780	25
Bromomethane	94.9	106	...	0.108	1.10	900	20
2-Butanone	72	0.0274	...	...	0.26	268,000	...
Carbon disulfide	76.1	12	260	...	2	2,940	20
Chlorobenzene	112.6	3.72	11.7	0.076	2.84	488	25
Chloroethane	64.5	14.8	...	...	1.54	5,740	25
Chloroform	119.4	3.39	160	0.094	1.97	8,000	20
Chloromethane	50.5	44	349	...	0.95	6,450	20
Dibromochloromethane	208.3	2.08	...	...	2.09	0.2	...
Dibromomethane	173.8	0.998	...	...	...	11,000	...
1,1-Dichloroethane	99.0	4.26	180	0.096	1.80	5,500	20
1,2-Dichloroethane	99.0	0.98	610	...	1.53	8,690	20
1,1-Dichloroethylene	97.0	34	600	0.084	1.84	210	25
1,2-Dichloroethylene	96.9	606	208	...	0.48	600	20
1,2-Dichloropropane	113.0	2.31	42	...	2.00	2,700	20
1,3-Dichloropropylene	111.0	3.55	380	...	1.98	2,800	25
Ethyl benzene	106.2	6.44	7	0.071	3.15	152	20
Methylene chloride	84.9	2.03	349	...	1.3	16,700	25
Pyrene	202.3	0.005	...	...	4.88	0.16	26
Styrene	104.1	9.7	5.12	0.075	2.95	300	20
1,1,1,2-Tetrachloroethane	167.8	0.381	5	0.077	3.04	200	20
1,1,2,2-Tetrachloroethane	167.8	0.38	...	...	2.39	2,900	20
Tetrachloroethylene	165.8	25.9	...	0.077	2.6	150	20
Tetrachloromethane	153.8	230	...	...	2.64	785	20
Toluene	92.1	6.7	22	0.083	2.73	515	20
Tribromoethane	252.8	0.552	5.6	...	2.4	3,200	30
1,1,1-Trichloroethane	133.4	14.4	100	...	2.49	4,400	20
1,1,2-Trichloroethane	133.4	1.17	...	...	2.47	4,500	20
Trichloroethylene	131.4	9.1	60	...	2.38	1,100	25
Trichlorofluoromethane	137.4	58	667	0.083	2.53	1,100	25
Vinyl chloride	62.5	81.9	2660	0.114	1.38	1.1	25
Xylenes	106.2	5.1	10	0.076	3.0	198	20

Source: [4, 11].

Note: The temperatures listed are the temperatures for the solubility values.

Table 2.5 summarizes the values of Henry's constant, vapor pressure, and solubility of some commonly encountered COCs. The values of the octanol–water partition coefficient, K_{ow}, and the diffusion coefficients, D, are also listed, and discussion on these two parameters will be given in later sections. For more complete lists of these values, chemistry handbooks and references should be consulted.

Example 2.28: Unit Conversions for Henry's Constant

As shown in Table 2.5, the Henry's constant for benzene in water at 25°C is 5.55 atm/M. Convert this value to dimensionless units and also to units of atm.

Solution:

From Table 2.4

$$H = H^*RT = 5.55 = H^*(0.082)(273 + 25)$$
$$H^* = 0.227 \text{ (dimensionless)}$$

Also, from Table 2.4

$$H = (H^*RT)[1,000\gamma/W]$$
$$= [(0.227)(0.082)(273 + 25)][(1,000)(1)/(18)] = 308.3 \text{ atm}$$

Discussion:

1. As mentioned previously, use of the dimensionless Henry's constant is becoming more popular. Benzene is a VOC of concern and is shown in most, if not all, databases of Henry's constant values. It may not be a bad idea to memorize that benzene has a dimensionless Henry's constant of 0.23 under ambient conditions.

2. To convert the Henry's constant of another COC in the database, just multiply the ratio of the Henry's constants (in any units) of that COC and of benzene by 0.23. For example, to find the dimensionless Henry's constant of methylene chloride, first read the Henry's constant for methylene chloride, 2.03 atm/M, and for benzene, 5.55 atm/M, from Table 2.5. Then find the ratio of these two and multiply it by 0.23, as $[(2.03)/(5.55)] \times (0.23) = 0.084$.

Example 2.29: Estimate Henry's Constant from
Solubility and Vapor Pressure

As shown in Table 2.5, the vapor pressure of benzene is 95.2 mm-Hg, and its solubility in water is 1,780 mg/L at 25°C. Estimate the Henry's constant of benzene from the given information.

Solution:

From Equation 2.21, we know that Henry's constant is the ratio of vapor pressure and solubility, so

$$H = (95.2 \text{ mm-Hg}) \div (1,780 \text{ mg/L}) = 0.0535 \text{ mm-Hg/(mg/L)}$$

To compare with the value given in Table 2.5, we need to do some conversions of the units:

$$P^{vap} = 95.2/760 = 0.125 \text{ atm}$$
$$C = 1{,}780 \text{ mg/L} = 1.78 \text{ g/L}$$
$$= (1.78 \text{ g/L}) \div (78.1 \text{ g/g-mole}) = 0.0228 \text{ mole/L} = 0.0228 \text{ M}$$
$$\text{So, } H = (0.125 \text{ atm}) \div (0.0228 \text{ M}) = 5.48 \text{ atm/M}$$

Discussion:

1. The calculated value, 5.48, is essentially the same as the value in Table 2.5, i.e., 5.55.
2. It should be noted that values of vapor pressure, solubility, and Henry's constant mentioned in a technical article might come from different sources. So the Henry's constant derived from the ratio of vapor pressure and solubility might not match well with the stated value of the Henry's constant.

Example 2.30: Use Henry's Law to Calculate the Equilibrium Concentrations

The subsurface of a site is impacted by tetrachloroethylene (PCE). A recent soil vapor survey indicates that the soil vapor contained 1,250 ppmV of PCE. Estimate the PCE concentration in the soil moisture. Assume the subsurface temperature is equal to 20°C.

Solution:

(a) From Table 2.5, for PCE

$$H = 25.9 \text{ atm/M and MW} = 165.8$$

Also, $1{,}250 \text{ ppmV} = 1{,}250 \times 10^{-6} \text{ atm} = 1.25 \times 10^{-3} \text{ atm} = P_A$

Use Equation (2.19):

$$P_A = H_A C_A$$
$$= 1.25 \times 10^{-3} \text{ atm} = (25.9 \text{ atm/M}) \times (C_A)$$
$$\text{So, } C_A = (1.25 \times 10^{-3}) \div 25.9$$
$$= 4.82 \times 10^{-5} \text{ M} = (4.82 \times 10^{-5} \text{ mole/L})(165.8 \text{ g/mole})$$
$$= 8 \times 10^{-3} \text{ g/L} = 8 \text{ mg/L} = 8 \text{ ppm}$$

(b) We can also use the dimensionless Henry's constant to solve this problem.

$$H = H^*RT = 25.9 = H^*(0.082)(273 + 20)$$
$$H^* = 1.08 \text{ (dimensionless)}$$

Use Equation (2.1) to convert ppmV to mg/m³:

$$1{,}250 \text{ ppmV} = (1{,}250)[(165.8/24.05)] \text{ mg/m}^3$$
$$= 8{,}620 \text{ mg/m}^3 = 8.62 \text{ mg/L}$$

Use Equation (2.21):

$$G = HC$$
$$8.62 \text{ mg/L} = (1.08)C$$
$$\text{So, } C = 8 \text{ mg/L} = 8 \text{ ppm}$$

Discussion:

1. These two approaches yield identical results.
2. Henry's constant of PCE is relatively high (five times higher than that of benzene, 1.08 vs. 0.227).
3. A concentration of 8 mg/L of PCE in soil moisture is in equilibrium with a vapor concentration of 1,250 ppmV.
4. The numeric value of the vapor concentration (1,250 ppm) is much higher than that of the corresponding liquid concentration (8 ppm).

2.4.3 Solid–Liquid Equilibrium

Adsorption. Adsorption is the process in which a compound moves from liquid phase onto the surface of the solid across the interfacial boundary. Adsorption is caused by interactions among three distinct components:

- Adsorbent (e.g., vadose zone soil, aquifer matrix, and activated carbon)
- Adsorbate (e.g., the COC)
- Solvent (e.g., soil moisture and groundwater)

In adsorption, the adsorbate is removed from the solvent and taken by the adsorbent. Adsorption is an important mechanism governing the COC's fate and transport in the environment.

Adsorption Isotherms. For a system where solid phase and liquid phase coexist, an adsorption isotherm describes the equilibrium relationship between the liquid and solid phases. The isotherm indicates that the relationship is for a constant temperature.

The most popular isotherms are the Langmuir isotherm and the Freundlich isotherm. Both were derived in the early 1900s. The Langmuir isotherm has a theoretical basis that assumes monolayer coverage of the adsorbent surface by the adsorbates, while the Freundlich isotherm is a semi-empirical relationship. For a Langmuir isotherm, the concentration on the soil increases with increasing concentration in liquid until a maximum concentration on the solid is reached. The Langmuir isotherm can be expressed as follows:

$$S = S_{\max} \frac{KC}{1 + KC} \qquad (2.22)$$

where S is the adsorbed concentration on the solid surface, C is the dissolved concentration in liquid, K is the equilibrium constant, and $S_{\max}$ is the maximum adsorbed concentration. It should be noted that, in this book, both symbols S and X are used for COC concentration in soil. Symbol S means "mass of COC/mass of dry soil," while X means "mass of COC/mass of soil plus moisture."

On the other hand, the Freundlich isotherm can be expressed in the following form:

$$S = KC^{1/n} \tag{2.23}$$

where both K and $1/n$ are empirical constants. These constants are different for different adsorbates, adsorbents, and solvents. For a given compound, the values will also be different for different temperatures. When using the isotherms, we should ensure that the units among the parameters and the empirical constants are consistent.

Both isotherms are nonlinear. Incorporating the nonlinear Langmuir or Freundlich isotherm into the mass balance equation to evaluate the COC's fate and transport will make the computer simulation more difficult or more time consuming. Fortunately, it was found that, in many environmental applications, the linear form of the Freundlich isotherm applies. It is called the linear adsorption isotherm, since $1/n = 1$, thus

$$S = KC \tag{2.24}$$

which simplifies the mass balance equation in a fate and transport model.

Partition Coefficient. For soil–water systems, the linear adsorption isotherm is often written in the following form:

$$S = K_p C$$

thus

$$K_p = S/C \tag{2.25}$$

where K_p is called the *partition coefficient* that measures the tendency of a compound to be adsorbed onto the surface of soil or sediment from a liquid phase. It describes how a COC distributes (partitions) itself between the two media (i.e., solid and liquid). Henry's constant, which was discussed previously, can be viewed as the vapor–liquid partition coefficient.

For a given organic chemical compound, the partition coefficient is not the same for every soil. The dominant mechanism of organic adsorption is the hydrophobic bonding between the compound and the natural organics associated with the soil. It was found that K_p increases linearly with the fraction of organic carbon (f_{oc}) in soil, thus

$$K_p = f_{oc} K_{oc} \tag{2.26}$$

The organic carbon partition coefficient (K_{oc}) can be considered as the partition coefficient for the organic compound into a hypothetical pure organic carbon phase. For soil that is not 100% organics, the partition coefficient is discounted by the factor, f_{oc}. Clayey soil is often associated with more natural organic matter and, thus, has a stronger adsorption potential for organic COCs.

K_{oc} is actually a theoretical parameter, and it is the slope of experimentally determined K_p vs. f_{oc} curves. K_{oc} values for many compounds are not readily available. Much research has been conducted to relate them to more commonly available chemical properties such as solubility in water (S_w) and the octanol–water partition coefficient (K_{ow}). The octanol–water partition coefficient is a dimensionless constant defined by:

$$K_{ow} = \frac{C_{octanol}}{C_{water}} \tag{2.27}$$

where

$C_{octanol}$ = equilibrium concentration of an organic compound in octanol
C_{water} = equilibrium concentration of the organic compound in water

K_{ow} serves as an indicator of how an organic compound will partition between an organic phase and water. Values of K_{ow} range widely, from 10^{-3} to 10^7. Organic compounds with low K_{ow} values are hydrophilic (like to stay in water) and have low soil adsorption. There are many correlation equations between K_{oc} and K_{ow} (or solubility in water, S_w) reported in the literature. Table 2.6 lists the ones mentioned in an EPA handbook (EPA 1991). It can be seen that K_{oc} increases linearly with increasing K_{ow} or with decreasing S_w on a log-log plot. (Note: Values of K_{ow} for some commonly encountered COCs are provided in Table 2.5.) The following simple correlation is also commonly used [4]:

$$K_{oc} = 0.63 K_{ow} \tag{2.28}$$

TABLE 2.6

Some Correlation Equations between K_{oc} and K_{ow} (or S_w)

Equation	Database
$\log K_{oc} = 0.544 (\log K_{ow}) + 1.377$ or $\log K_{oc} = -0.55 (\log S_w) + 3.64$	Aromatics, carboxylic acids and esters, insecticides, ureas and uracils, triazines, miscellaneous
$\log K_{oc} = 1.00 (\log K_{ow}) - 0.21$	Polycyclic aromatics, chlorinated hydrocarbons
$\log K_{oc} = -0.56 (\log S_w) + 0.93$	PCBs, pesticides, halogenated ethanes and propanes, PCE, 1,2-dichlorobenzene

Source: [12].
Note: S_w is the solubility in water, in mg/L.

To estimate the solid concentration in equilibrium with the liquid concentration (or vice versa), we have to determine the value of the partition coefficient first. The following procedure can be used to determine the partition coefficient for a soil–water system:

Step 1: Find K_{ow} or S_w of the COC (Table 2.5).

Step 2: Determine K_{oc} using correlations given in Table 2.6 or Equation (2.28).

Step 3: Determine f_{oc} of the soil.

Step 4: Determine K_p using Equation (2.26).

Example 2.31: Solid–Liquid Equilibrium Concentrations

The aquifer underneath a site is impacted by tetrachloroethylene (PCE). A groundwater sample contains 200 ppb of PCE. Estimate the PCE concentration adsorbed onto the aquifer material, which contains 1% of organic carbon. Assume the adsorption isotherm follows a linear model.

Solution:

(a) From Table 2.5, for PCE
$$\log K_{ow} = 2.6 \rightarrow K_{ow} = 398$$

(b) From Table 2.6, for PCE (a chlorinated hydrocarbon)
$$\log K_{oc} = 1.00(\log K_{ow}) - 0.21$$
$$= 2.6 - 0.21 = 2.39$$
$$K_{oc} = 245 \text{ mL/g} = 245 \text{ L/kg}$$
Or, from Equation (2.28)
$$K_{oc} = 0.63 K_{ow} = 0.63(398) = 251 \text{ mL/g} = 251 \text{ L/kg}$$

(c) Use Equation (2.26) to find K_p:
$$K_p = f_{oc} K_{oc}$$
$$= (1\%)(251) = 2.51 \text{ mL/g} = 2.51 \text{ L/kg}$$

(d) Use Equation (2.25) to find S:
$$S = K_p C$$
$$= (2.51 \text{ L/kg})(0.2 \text{ mg/L}) = 0.50 \text{ mg/kg}$$

Discussion:

1. Equation (2.28) ($K_{oc} = 0.63 K_{ow}$), which looks very simple, yields an estimate of K_{oc} (251 kg/L) that is comparable to the value (245 L/kg) from using the correlation equation in Table 2.6.

2. Most technical articles do not talk about the units of K_p. Actually, K_p has a unit of "(volume of solvent)/(mass of adsorbent)," and it is equal to mL/g or L/kg in most, if not all, of the correlation equations.

2.4.4 Solid–Liquid–Vapor Equilibrium

As mentioned in the beginning of this section, an NAPL may end up in four different phases as it enters a vadose zone. We have just discussed the equilibrium systems of liquid–vapor and soil–liquid. Now we move one step further to discuss the system including liquid, vapor, and solid (and free product in some of the applications).

The soil moisture in the vadose zone is in contact with both soil grains and air in the void, and the COC in each phase can travel to the other phases. The dissolved concentration in the liquid, for example, is affected by the concentrations in the other phases (i.e., soil, vapor, and free product). If the entire system is in equilibrium, these concentrations are related by the equilibrium equations mentioned previously. In other words, if the entire system is in equilibrium and the COC concentration of one phase is known, the concentrations at other phases can be estimated using the equilibrium relationships. Although, in real applications, the equilibrium condition does not always exist, the estimate from such a condition serves as a good starting point or as the upper or the lower limit of the real values.

Example 2.32: Solid–Liquid–Vapor-Free-Product Equilibrium Concentrations

Free-product phase of 1,1,1-trichloroethane (1,1,1-TCA) was found in the subsurface at a site. The soil is silty, with an organic content of 2%. The subsurface temperature is 20°C. Estimate the maximum concentrations of TCA in the air void, in the soil moisture, and on the soil grains.

Solution:

(a) Since the free product is present, the maximum vapor concentration will be the vapor pressure of the TCA liquid at that temperature.

From Table 2.5, the vapor pressure of TCA is 100 mm-Hg at 20°C.

$$100 \text{ mm-Hg} = (100 \text{ mm-Hg}) \div (760 \text{ mm-Hg/atm}) = 0.132 \text{ atm}$$
$$G = 0.132 \text{ atm} = 132,000 \text{ ppmV}$$

Use Equation (2.1) to convert ppmV to mg/m³ (MW = 133.4 from Table 2.5)

$$132,000 \text{ ppmV} = (132,000)(133.4/24.05) \text{ mg/m}^3$$
$$G = 732,200 \text{ mg/m}^3 = 732.2 \text{ mg/L}$$

(b)   From Table 2.5, $H = 14.4$

Convert H to dimensionless Henry's constant, using Table 2.4:
$$H = H^*RT = 14.4 = H^*(0.082)(273 + 20)$$
$$H^* = 0.60 \text{ (dimensionless)}$$
Use Equation (2.20) to find the liquid concentration:
$$G = HC = 732.2 \text{ mg/L} = (0.60)C$$
So, $C = 1{,}220 \text{ mg/L} = 1{,}220 \text{ ppm}$

(c)   From Table 2.5, for TCA
$$\log K_{ow} = 2.49 \rightarrow K_{ow} = 309$$
From Table 2.6, for TCA (a chlorinated hydrocarbon)
$$\log K_{oc} = 1.00(\log K_{ow}) - 0.21$$
$$= 2.49 - 0.21 = 2.28$$
$$K_{oc} = 191 \text{ mL/g} = 191 \text{ L/kg}$$
Or, from Equation (2.28)
$$K_{oc} = 0.63 \, K_{ow} = 0.63(309) = 195 \text{ mL/g} = 195 \text{ L/kg}$$
Use Equation (2.26) to find K_p:
$$K_p = f_{oc} K_{oc}$$
$$= (2\%)(191) = 3.82 \text{ mL/g} = 3.82 \text{ L/kg}$$
Use Equation (2.25) to find the soil concentration, S:
$$S = K_p C$$
$$= (3.82 \text{ L/kg})(1{,}220 \text{ mg/L}) = 4{,}660 \text{ mg/kg}$$

Discussion:

1. The calculated liquid concentration, 1,220 mg/L, is smaller than the solubility, 4,400 mg/L, as given in Table 2.5.
2. The simple equation, Equation (2.28), again yields an estimate of K_{oc} (195 L/kg) that is comparable to the value (191 L/kg) from using the correlation equation in Table 2.6.
3. The calculated concentrations are the maximum possible values; the actual values would be lower if the system is not in equilibrium and not a confined system.

Example 2.33: Solid–Liquid–Vapor Equilibrium Concentrations (Absence of Free Product)

For a subsurface impacted by 1,1,1-trichloroethane (1,1,1-TCA), the soil vapor concentration at a location was found to be 1,320 ppmV. The soil is silty, with an organic content of 2%. The subsurface temperature is 20°C. Estimate the maximum concentrations of TCA in the soil moisture and on the soil grains.

Solution:

(a) With a concentration of 1,320 ppmV, this is 100 times smaller than that in Example 2.32:

$$G = 1{,}320 \text{ ppmV} = 7{,}320 \text{ mg/m}^3 = 7.32 \text{ mg/L}$$

(b) With dimensionless Henry's constant of 0.60 (from Example 2.32):

$$G = HC = 7.32 \text{ mg/L} = (0.60)C$$

So, $C = 12.2$ mg/L = 12.2 ppm

(c) With $K_p = 3.82$ L/kg,

$$S = K_p C = (3.82 \text{ L/kg})(12.2 \text{ mg/L}) = 46.6 \text{ mg/kg}$$

Discussion:

1. The equilibrium relationships ($G = HC$ and $S = K_p C$) are linear. With a vapor concentration of 1,320 ppmV, which is 100 times smaller than that (132,000 ppmV) in Example 2.32, the corresponding liquid and solid concentrations are correspondingly smaller by 100 times. It should be noted that this is only valid when two systems have the same characteristics (i.e., same H and K_p values).

2. The concentrations are based on an assumption that the system is in equilibrium; the actual values would be different if the system is not in equilibrium.

2.4.5 Partition of COCs in Different Phases

The total mass of COCs in the vadose zone is the sum of the mass in four phases (vapor, moisture, solid, and free product). Let us consider a COC plume in the vadose zone with a volume, V.

From Equation (2.5),

$$\text{mass of COC dissolved in the soil moisture} = (V_l)(C) = [V(\phi_w)]C \quad (2.29)$$

From Equation (2.6),

$$\text{mass of COC adsorbed onto the soil grains} = (M_s)(S) = [V(\rho_b)]S \quad (2.30)$$

From Equation (2.7),

$$\text{mass of COC in the void space} = (V_a)(G) = [(V)(\phi_a)]G \quad (2.31)$$

where ϕ_w is the volumetric water content and ϕ_a is the air porosity. (Note: total porosity, $\phi_t = \phi_w + \phi_a$.) The total mass of COC (M_t) present in the plume is the sum of the mass in the previously mentioned three phases and free product, if any. Thus,

$$M_t = V(\phi_w)C + (V)(\rho_b)S + V(\phi_a)G + \text{mass of the free product} \qquad (2.32)$$

The mass of free product is simply the volume of the free product multiplied by its mass density. If no free product is present, Equation (2.32) can be simplified to:

$$M_t = V(\phi_w)C + (V)(\rho_b)S + (V)(\phi_a)G \qquad (2.33)$$

If the system is in equilibrium and both Henry's law and the linear adsorption apply, the concentration in one phase can be represented by the concentration in another phase multiplied by a factor. The following relationships exist:

$$G = HC = H\left(\frac{S}{K_p}\right) = \left(\frac{H}{K_p}\right)S \qquad (2.34)$$

$$C = \left(\frac{S}{K_p}\right) = \left(\frac{G}{H}\right) \qquad (2.35)$$

$$S = K_pC = K_p\left(\frac{G}{H}\right) = \left(\frac{K_p}{H}\right)G \qquad (2.36)$$

Using these relationships, Equation (2.33) can be rearranged to:

$$\begin{aligned}\frac{M_t}{V} &= [(\phi_w) + (\rho_b)K_p + (\phi_a)H]C \\ &= \left[\frac{(\phi_w)}{H} + \frac{(\rho_b)K_p}{H} + (\phi_a)\right]G \\ &= \left[\frac{(\phi_w)}{K_p} + \rho_b + (\phi_a)\frac{H}{K_p}\right]S\end{aligned} \qquad (2.37)$$

where M_t/V can be viewed as the average mass concentration of the plume. The total mass of COCs in a plume can be readily determined by multiplying

(M_t/V), if known, with the total volume of the plume. Equation (2.37) can be used to estimate the total mass of COC in a vadose zone if the average soil moisture concentration, soil concentration, or vapor concentration is known, and if no free product is present.

For a dissolved groundwater plume ($\phi_a = 0$ and $\phi_w = \phi_t$), Equation (2.37) can be modified to

$$\frac{M_t}{V} = [\phi + (\rho_b)K_p]C$$

$$= \left[\frac{\phi}{K_p} + \rho_b\right]S \qquad (2.38)$$

To use the equations in this subsection, the following units are suggested: V (in liters), G (mg/L), C (mg/L), S (mg/kg), M_t (mg), ρ_b (kg/L), K_p (L/kg), and ϕ_t, ϕ_w, ϕ_a, and H (dimensionless).

As mentioned in Section 2.4.3, both S and X are used for COC concentrations in soil in this book. S is the adsorbed concentration on the solid surface, and X is used to represent the COC concentration of a soil sample. Symbol S means "mass of COC/mass of dry soil," while X means "mass of COC/mass of soil plus moisture." Assuming the mass of COC in the void is also captured in the analysis of the soil sample, the total mass of COC contained in a unit volume of soil (M_t/V) can be related to COC concentration in soil (X) as

$$\frac{M_t}{V} = X \times \rho_t \qquad (2.39)$$

where ρ_t is the total bulk density of the soil sample

As will be shown in Example 2.37, the mass in the void is relatively small compared to that in the dissolved and adsorbed phases. Consequently, the inclusion of mass in the void in Equation (2.39) is acceptable. Inserting Equation (2.39) into Equation (2.37), the soil sample concentration (X) can be related to G, C, and S as

$$X = \left\{\frac{[(\phi_w) + (\rho_b)K_p + (\phi_a)H]}{\rho_t}\right\} \times C$$

$$= \left\{\frac{\left[\frac{(\phi_w)}{H} + \frac{(\rho_b)K_p}{H} + (\phi_a)\right]}{\rho_t}\right\} \times G \qquad (2.40)$$

$$= \left\{\frac{\left[\frac{(\phi_w)}{K_p} + \rho_b + (\phi_a)\frac{H}{K_p}\right]}{\rho_t}\right\} \times S$$

Example 2.34: Mass Partition between Vapor and Liquid Phases

A new field technician was sent out to collect a groundwater sample from a monitoring well. He filled only half of the 40-mL sample vial with groundwater impacted by benzene ($T = 20°C$). The benzene concentration in the collected groundwater was analyzed to be 5 mg/L.

Determine:

(a) The concentration of benzene in the head space (in ppmV) before the vial was opened

(b) The percentage of total benzene mass in the aqueous phase of the closed vial

(c) The true benzene concentration in the groundwater, if headspace free sample is collected

Assume the value of the dimensionless Henry's constant for benzene is equal to 0.22.

Solution:

Basis: 1-liter container

(a) Concentration of benzene in the headspace
$$= (H)(C)$$
$$= (0.22)(5) = 1.1 \text{ mg/L} = 1{,}100 \text{ mg/m}^3$$
$$1 \text{ ppmV} = (MW/24.05) \text{ mg/m}^3 = (78/24.05) \text{ mg/m}^3 = 3.24 \text{ mg/m}^3$$
Concentration of benzene in the head space
$$= 1{,}100/3.24 = 340 \text{ ppmV}$$

(b) Mass of benzene in the liquid phase
$$= (C)(\text{volume of the liquid})$$
$$= (5)(1 \times 50\%) = 2.5 \text{ mg}$$
Mass of benzene in the headspace
$$= (G)(\text{volume of the air space})$$
$$= (1.1)(1 - 0.5) = 0.55 \text{ mg}$$
Total mass of benzene
$$= \text{mass in the liquid} + \text{mass in the headspace}$$
$$= 2.5 + 0.55 = 3.05 \text{ mg}$$
Percentage of total benzene mass in the aqueous phase
$$= 2.5/3.05 = 82\%$$

(c) The actual liquid concentration should be
$$= (\text{total mass of benzene}) \div (\text{volume of the liquid})$$
$$= (3.05)/(0.5) = 6.1 \text{ mg/L}$$

Discussion:

1. Although the sample volume is only 40 mL, the calculation basis was 1 L to simplify the calculation.
2. With the presence of the headspace in the sample bottle, the apparent liquid concentration was lower than the actual concentration.

Example 2.35: Mass Partition between Solid and Liquid Phases in an Aquifer

The aquifer underneath a site is impacted by tetrachloroethylene (PCE). The aquifer porosity is 0.4, and the (dry) bulk density of the aquifer material is 1.6 g/cm^3. A groundwater sample contains 200 ppb of PCE.

Assuming that the adsorption follows a linear model, estimate:

(a) The PCE concentration adsorbed on the aquifer material, which contains 1% by weight of organic carbon.
(b) The partition of PCE in two phases, i.e., dissolved phase and adsorbed onto the solid phase.

Solution:

(a) The PCE concentration adsorbed onto the solid has been determined in Example 2.31 as 0.50 mg/kg.

(b) Basis: 1-L aquifer formation

Mass of PCE in the liquid phase

$$= (C)[(V)(\phi)] = (0.2)[(1)(0.4)] = 0.08 \text{ mg}$$

Mass of PCE adsorbed on the solid

$$= (S)[(V)(\rho_b)] = (0.5)[(1)(1.6)] = 0.8 \text{ mg}$$

Total mass of PCE = mass in the liquid + mass on the solid

$$= 0.08 + 0.8 = 0.88 \text{ mg}$$

Percentage of total PCE mass in the aqueous phase

$$= 0.08/0.88 = 9.1\%$$

Discussion:

Most of the PCE, 90.9%, in the impacted aquifer is adsorbed onto the aquifer materials. This partially explains why the cleanup of aquifers takes a long time using the pump-and-treat method.

Example 2.36: Mass Partition between Liquid and Solid Phases

A wastewater contains 500 mg/L of suspended solids. The fraction of organics of the solids is 1% by weight. The benzene concentration of the *filtered* wastewater is determined to be 5 mg/L. K_{oc} of benzene is 85 mL/g.

Determine:

(a) The concentration of benzene adsorbed onto the suspended solids, and

(b) The percentage of total benzene mass in the dissolved phase of the unfiltered wastewater.

Solution:

(a) Use Equation (2.26) to find K_p:

$$K_p = f_{oc} K_{oc}$$
$$= (1\%)(85) = 0.85 \text{ mL/g} = 0.85 \text{ L/kg}$$

Use Equation (2.25) to find S:

$$S = K_p C$$
$$= (0.85 \text{ L/kg})(5 \text{ mg/L}) = 4.25 \text{ mg/kg}$$

(b) Basis: 1-L solution

Mass of benzene in the liquid phase

$$= (C)[(\text{volume of the solution})] = (5)(1) = 5 \text{ mg}$$

Mass of benzene adsorbed on the solid

$$= (S)[(\text{Volume of the solution})(\text{suspended solid concentration})]$$
$$= (4.15 \text{ mg/kg})[(1 \text{ L})(5{,}000 \text{ mg/L})(1 \text{ kg/1{,}000{,}000 mg}]$$
$$= 2.125 \times 10^{-2} \text{ mg}$$

Total mass of benzene

$$= \text{mass in the liquid} + \text{mass on the solid}$$
$$= 5 + (2.125 \times 10^{-2}) = 5.021 \text{ mg}$$

Percentage of total benzene mass in the aqueous phase

$$= 5/(5.0215) = 99.6\%$$

Discussion:

Almost all of the benzene, 99.6%, is in the dissolved phase because only a small amount of solids is present, the organic content of the suspended solids is low, and benzene is not very hydrophobic.

Example 2.37: Mass Partition between Vapor, Liquid, and Solid Phases

The vapor concentrations of benzene and pyrene in the void space of the vadose zone underneath a landfill are 100 ppmV and 10 ppbV, respectively. The total porosity of the vadose zone is 40%, and 30% of the porosity is occupied by water. The (dry) bulk density of the soil is 1.6 g/cm³, and the total bulk density is 1.8 g/cm³. Assuming no free product is present, determine the mass fractions of each COC in the three phases (i.e., void, moisture, and solid phases). The values of the dimensionless Henry's constant for benzene and pyrene are 0.22 and 0.0002, respectively. The values of K_p for benzene and pyrene are 1.28 and 717, respectively.

Strategy:

A computer spreadsheet, such as Microsoft Excel, is a good way to solve a problem such as this.

Solution:

Basis: 1 m³ of soil

(a)   Determine the mass in the void

	Benzene	Pyrene
MW	78	202
G (ppmV)	100	0.01
G (mg/m³)	324.3	0.084
Air void (m³) = 0.40×0.7	0.28	0.28
Mass in void (mg)	90.8	0.024

(b)   Determine the mass dissolved in the liquid

	Benzene	Pyrene
H	0.22	0.0002
C (mg/m³) = G/H	1,474	420
Liquid volume (m³) = 0.40×0.3	0.12	0.12
Mass in liquid (mg)	176.9	50.4

(c)   Determine the mass adsorbed onto the solid

	Benzene	Pyrene
K_p	1.28	717
C (mg/L)	1.47	0.42
S (mg/kg) = $K_p \times C$ (mg/L)	1.89	301
Soil mass (kg) = $(1\ m^3)(\rho_b)$	1,600	1,600
Mass on solid surface (mg)	3,019	4.82×10^5

(d)   Determine the total mass in three phases

	Benzene	Pyrene
Total COC (mg)	3,287	4.82×10^5

(e)   Determine the mass fraction in each phase

	Benzene	Pyrene
% in void	2.8	4.9×10^{-6}
% in moisture	5.4	0.01
% in solid	91.8	99.99

Discussion:

For both compounds, most of the COCs are attached onto the solid (91.8% for benzene and 99.99% for pyrene). This is especially true for pyrene that has very high K_p and low H values. The vapor concentration of pyrene is extremely low, while its concentration in soil is very high.

Example 2.38: COC Concentrations in Soil: A Comparison of S and X

In Example 2.37, the concentrations of benzene and pyrene adsorbed onto the soil grains were found to be 1.89 and 301 mg/kg, respectively. If a sample was taken from that location and analyzed for benzene and pyrene concentration in soil by a laboratory, what would be the concentration values of the soil samples? Use these values to estimate the total COC mass in soil.

Solution:

(a)   The values of X and S are related by Equation (2.40)

$$X = \left\{ \frac{\left[\frac{(\phi_w)}{K_p} + \rho_b + (\phi_a)\frac{H}{K_p} \right]}{\rho_t} \right\} \times S$$

$X = \{[(0.12)/(1.28) + 1.6 + (0.28)(0.22)/(1.28)] \div 1.8\} \times 1.89$
$\quad = 1.83$ mg/kg (for benzene)
$X = \{[(0.12)/(717) + 1.6 + (0.28)(0.0002)/(717)] \div 1.8\} \times 301$
$\quad = 268$ mg/kg (for pyrene)

(b)   If the COC mass in the void was not captured in the laboratory analysis, Equation (2.40) can be modified to:

$$X = \left\{ \frac{\left[\frac{(\phi_w)}{K_p} + \rho_b \right]}{\rho_t} \right\} \times S$$

$X = \{[(0.12)/(1.28) + 1.6] \div 1.8\} \times 1.89$
$\quad = 1.78$ mg/kg (for benzene)
$X = \{[(0.12)/(717) + 1.6] \div 1.8\} \times 301$
$\quad = 268$ mg/kg (for pyrene)

(c) The total mass of the COC in soil can be found using Equation (2.7) as

$$(V_s)(\rho_t)](X) = (M_s)(X)$$
$$= [(1 \text{ m}^3)(1{,}800 \text{ kg/m}^3)](1.83 \text{ mg/kg})$$
$$= 3{,}284 \text{ mg (for benzene)}$$
$$= [(1 \text{ m}^3)(1{,}800 \text{ kg/m}^3)](268 \text{ mg/kg})$$
$$= 482{,}000 \text{ mg (for pyrene)}$$

Discussion:

1. This example illustrates the differences between X and S. For benzene, the values of X and S are relatively close. For pyrene, the ratio of the X and S values is essentially the ratio of the dry bulk density and total bulk density, mainly because the majority of the pyrene compounds are adsorbed on the surface of the soil grains.

2. Neglecting the mass in the void has an insignificant impact on the estimated values of X.

3. The calculated values of total mass in soil are essentially the same as those in Example 2.37.

Example 2.39: Relationship between Soil Vapor Concentration and Soil Sample Concentration

The vapor concentrations of benzene and pyrene in the void space of the vadose zone underneath a landfill are 100 ppmV and 10 ppbV, respectively, from a soil gas survey. The total porosity of the vadose zone is 40%, and 30% of the void is occupied by water. The (dry) bulk density of the soil is 1.6 g/cm^3 and the total bulk density is 1.8 g/cm^3. The values of the dimensionless Henry's constant for benzene and pyrene are 0.22 and 0.0002, respectively. Values of K_p for benzene and pyrene are 1.28 and 717, respectively.

Soil samples were taken from the location where the soil gas probe was located and then analyzed in a laboratory for the COC concentrations in soil. Estimate the COC concentrations in soil.

Solution:

Basis: 1 L of soil

(a) Let us work on benzene first. We have to convert the vapor concentration in ppmV into mg/L first. From Example 2.37, $G = 0.324$ mg/L for benzene.

Use Equation (2.37) to estimate the mass concentration of benzene in the soil:

$$\frac{M_t}{V} = \left[\frac{(\phi_w)}{H} + \frac{(\rho_b)K_p}{H} + (\phi_a)\right]G$$

$$= \left[\frac{0.12}{0.22} + \frac{(1.6)(1.28)}{0.22} + 0.28\right](0.324) = 3.28 \text{ mg/L}$$

To convert this mass concentration in soil into mg/kg, we should divide the value by the total bulk density of the soil:

Soil concentration (X) = 3.28 mg/L ÷ 1.8 kg/L = 1.82 mg/kg (for benzene)

(b) For pyrene, from Example 2.37, $G = 0.000084$ mg/L.

Use Equation (2.37) to estimate the mass concentration of pyrene in soil:

$$\frac{M_t}{V} = \left[\frac{(\phi_w)}{H} + \frac{(\rho_b)K_p}{H} + (\phi_a)\right]G$$

$$= \left[\frac{0.12}{0.0002} + \frac{(1.6)(717)}{0.0002} + 0.28\right](0.000084) = 482 \text{ mg/L}$$

To convert this mass concentration in soil into mg/kg, we should divide the value by the total bulk density of the soil:

Soil concentration (X) = 482 mg/L ÷ 1.8 kg/L = 268 mg/kg (for pyrene)

Discussion:

1. In this example, a soil sample containing 1.82 mg/kg benzene yields a soil vapor concentration of 100 ppmV. The soil concentration of pyrene, 268 mg/kg, is 150 times larger than that of benzene, but its vapor concentration is 10,000 times smaller.

2. For a given COC concentration in soil, its soil vapor concentration will be higher if K_p value is smaller and the Henry's constant of the COC is larger. (In other words, the soil contains less organics and the COC is less hydrophobic and more volatile.) For sandy soil, the soil vapor concentration may be high, but the mass adsorbed onto the sand grains may be relatively low. This explains why PID or OVA readings on impacted sandy soil samples may be high; however, the laboratory results on the sandy soil samples might turn out to be very low.

3. The soil concentration of pyrene in this example, 268 mg/kg, means "268 mg pyrene per gram of dry soil plus soil moisture," while that in the previous example, 301 mg/kg, means "301 mg

pyrene per gram of dry soil." Typical laboratory results are on the basis of wet soil.

References

1. USEPA. 2010. Gasoline composition regulations affecting LUST sites. EPA 600/R-10/001. Washington, DC: Office of Research and Development, US Environmental Protection Agency.
2. USEPA. 2012. Summary and analysis of the 2011 gasoline benzene pre-compliance reports. EPA 420/R-12/007. Washington, DC: Office of Transportation and Air Quality, US Environmental Protection Agency.
3. US Federal Register. 2007. Control of hazardous air pollutants from mobile sources. 72 (37): 8427–76.
4. LaGrega, M.D., P.L. Buckingham, and J.C. Evans. 1994. *Hazardous waste management*. New York: McGraw-Hill.
5. Fetter Jr., C.W. 1980. *Applied hydrogeology*. Columbus, OH: Charles E. Merrill.
6. Todd, D.K. 1980. *Groundwater hydrology*. 2nd ed. New York: John Wiley & Sons.
7. Ballestero, T.P., F.R. Fiedler, and N.E. Kinner. 1994. An investigation of the relationship between actual and apparent gasoline thickness in a uniform sand aquifer. *Groundwater* 32 (5): 708.
8. Lide, D.R. 1992. *Handbook of chemistry and physics*. 73rd ed. Boca Raton, FL: CRC Press.
9. Reid, R.C., J.M. Prausnitz, and B.F. Poling. 1987. *The properties of liquids and gases*. 4th ed. New York: McGraw-Hill.
10. Kuo, J.F., and S.A. Cordery. 1988. Discussion of monograph for air stripping of VOC from water. *J. Environ. Eng.* 114 (5): 1248–50.
11. USEPA. 1990. CERCLA site discharges to POTWs treatability manual. EPA 540/2-90-007. Washington, DC: Office of Water, US Environmental Protection Agency.
12. USEPA. 1991. Site characterization for subsurface remediation. EPA 625/R-91/026. Washington, DC: US Environmental Protection Agency.
13. Agency for Toxic Substances & Disease Registry. 2005. *Public health assessment manual*. Appendix G: Calculating exposure doses (2005 update). ATSDR. http://www.atsdr.cdc.gov/hac/PHAManual/toc.html.

3

Plume Migration in Aquifer and Soil

3.1 Introduction

In Chapter 2, we illustrated necessary calculations for site assessment and remedial investigation. Generally, from remedial investigation (RI) activities, the extent of the contaminated plume in subsurface soil and/or aquifer would be defined. If the compounds of concern (COCs) are not removed, they may migrate farther under common field conditions, and the extent of the plume(s) will enlarge.

In the vadose zone, the COCs will move downward as free product and, in the meantime, get dissolved in infiltrating water and then move downward by gravity. The downward-moving liquid may come into contact with the underlying aquifer and create a dissolved plume. The dissolved plume will move downgradient in the aquifer. In addition, the COCs, especially the volatile organic compounds (VOCs), will volatilize into the air void of the vadose zone and travel under advective forces (with the air flow) and concentration gradients (through diffusion). Migration of the vapor can be in any direction, and the COCs in the vapor phase, when coming in contact with the soil moisture and groundwater, may get absorbed into them. For site remediation or health risk assessment, understanding the fate and transport of COCs in subsurface is important. Common questions related to the fate and transport of COCs in subsurface include:

1. How long will it take for the plume in the vadose zone to enter the aquifer?
2. How far and how fast will the vapor COCs in the vadose zone travel? In what concentrations?
3. How fast does the groundwater flow? In which direction?
4. How fast will the plume migrate? In which direction?
5. Will the plume migrate at the same speed of the groundwater flow? If different, what are the factors that would make the plume migrate at a different speed?
6. How long has the plume been present in the aquifer?

This chapter covers the basic calculations needed to answer most of these questions. The first section presents calculations for groundwater movement and clarifies some common misconceptions about groundwater velocity and hydraulic conductivity. The procedures to determine the groundwater flow gradient and flow direction are also given. The second section discusses groundwater extraction from confined and unconfined aquifers. Since hydraulic conductivity plays a pivotal role in groundwater movement, several common methodologies of estimating this parameter are covered, including the aquifer tests. The discussion then moves to the migration of the plume in the aquifer and in the vadose zone.

3.2 Groundwater Movement

3.2.1 Darcy's Law

Darcy's law is commonly used to describe laminar flow in porous media. For a given medium, the flow rate is proportional to the head loss and inversely proportional to the length of the flow path. Flow in typical groundwater aquifers is laminar, and therefore Darcy's law is valid. Darcy's law can be expressed as

$$v_d = \frac{Q}{A} = -K\frac{dh}{dl} \tag{3.1}$$

where v_d is the Darcy velocity, Q is the volumetric flow rate, A is the cross-sectional area of the porous medium perpendicular to the flow, dh/dl is the hydraulic gradient (a dimensionless quantity), and K is the hydraulic conductivity.

The hydraulic conductivity tells how permeable the porous medium is to the flowing fluid. The larger the K of a formation, the easier it is for the fluid to flow through it.

Commonly used units for hydraulic conductivity are either in velocity units such as ft/day, cm/s, or m/day, or in volumetric flow rate per unit area such as gpd/ft^2 or m^3/day/m^2. You may find the unit conversions in Table 3.1 helpful.

TABLE 3.1

Common Conversion Factors for Hydraulic Conductivity

m/day	cm/s	ft/day	gpd/ft^2
1	1.16×10^{-3}	3.28	2.45×10^{1}
8.64×10^{2}	1	2.83×10^{3}	2.12×10^{4}
3.05×10^{-1}	3.53×10^{-4}	1	7.48
4.1×10^{-2}	4.73×10^{-5}	1.34×10^{-1}	1

Example 3.1: Estimate the Rate of Groundwater Entering the Existing Plume

Leachates from a landfill leaked into the underlying aquifer and created a dissolved plume. Use the data below to estimate the amount of fresh groundwater that enters into the impacted zone per day:

- The maximum cross-sectional area of the plume perpendicular to the groundwater flow = 1,600 ft² (20′ in thickness × 80′ in width)

- Groundwater gradient = 0.005
- Hydraulic conductivity = 2,500 gpd/ft²

Solution:

Another common form of Darcy's law (Equation 3.1) is

$$Q = K \times (dh/dl) \times A = KiA \tag{3.2}$$

where i (= dh/dl) is the hydraulic gradient.

The rate of fresh groundwater entering the plume can be found by inserting the appropriate values into Equation (3.2):

$$Q = (2,500 \text{ gpd/ft}^2)(0.005)(1,600 \text{ ft}^2) = 20,000 \text{ gpd}$$

Discussion:

1. The calculation itself is straightforward and simple. However, we can get valuable and useful information from this exercise. The rate of 20,000 gallons per day represents the rate of upstream groundwater that will come into contact with the COCs. This water would become impacted and move downstream or sidestream and, consequently, enlarge the size of the plume.

2. To control the spread of the existing plume, one needs to extract this amount of water, 20,000 gpd (or ≈14 gallons per minute [gpm]), as the minimum. The actual extraction rate required should be larger than this, because the groundwater drawdown from pumping will increase the flow gradient. This increased gradient will, in turn, increase the rate of groundwater entering the impacted zone as indicated by Equation (3.2). In addition, not all the extracted water will come from the impacted zone

3. Using the maximum cross-sectional area is a legitimate approach that represents the "contact face" between the fresh groundwater and the impacted zone. The maximum cross-sectional areas could be found as the multiplication product of the maximum plume thickness and the maximum plume width.

3.2.2 Darcy Velocity versus Seepage Velocity

The velocity term in Equation (3.1) is often called the Darcy velocity (or the discharge velocity). Does the Darcy velocity represent the actual groundwater flow velocity? The straight answer to this question is "no." The Darcy velocity in Equation (3.1) assumes the flow occurs through the entire cross-section of the porous medium. In other words, it is the velocity with which water moves through an aquifer as if the aquifer were an open conduit. Actually, the flow is limited only to the available pore space (i.e., the effective cross-sectional area available for flow is smaller). Consequently, the actual fluid velocity through a porous medium would be larger than the corresponding Darcy velocity. This flow velocity is often called the seepage velocity or the interstitial velocity. The relationship between the seepage velocity, v_s, and the Darcy velocity, v_d, is

$$v_s = \frac{Q}{\phi A} = \frac{v_d}{\phi} \tag{3.3}$$

where ϕ is the effective porosity. For example, for an aquifer with an effective porosity of 33%, the seepage velocity will be three times that of the Darcy velocity (i.e., $v_s = 3 \, v_d$).

Example 3.2: Darcy Velocity versus Seepage Velocity

There was a spill of an inert substance into subsurface. The spill infiltrated the unsaturated zone and quickly reached the underlying water-table aquifer. The aquifer consists mainly of sand and gravel, with a hydraulic conductivity of 2,500 gpd/ft² and an effective porosity of 0.35. The static water level in a well neighboring the spill is 560 ft and that in another well, one mile directly downgradient, is 550 ft. Determine:

- The Darcy velocity of the groundwater
- The seepage velocity of the groundwater
- The velocity of the plume migration
- How long it will take for the plume to reach the downgradient well

Solution:

(a) We need to determine the hydraulic gradient first:

$$i = dh/dl$$
$$= (560 - 550)/(5,280) = 1.89 \times 10^{-3} \text{ ft/ft} = 1.89 \times 10^{-3}$$

Darcy velocity $(v_d) = Ki$

$$= \left[(2,500 \ \text{gpd/ft}^2) \left(0.134 \frac{\text{ft/day}}{\text{gpd/ft}^2} \right) \right] (1.89 \times 10^{-3}) = \underline{0.63 \ \text{ft/day}}$$

(b) Seepage velocity $(v_s) = v_d/\phi$
$$= 0.63/0.35 = 1.81 \text{ ft/day}$$

(c) The pollutant is inert, meaning that it will not react with the aquifer materials. (Sodium chloride is a good example as an inert substance, and it is one of the common tracers used in aquifer studies.) Therefore, the velocity of plume migration for this case is the same as the seepage velocity, 1.81 ft/day.

(d) Time = distance/velocity
$$= (5,280 \text{ ft}) \div (1.81 \text{ ft/day})$$
$$= 2,912 \text{ days} = 8.0 \text{ years}$$

Discussion:

1. The conversion factor (1 gpd/ft^2 = 0.134 ft/day), used in part (a), is from Table 3.1.

2. The calculated plume migration velocity is crude at best, and it should only be considered as a rough estimate. Many factors, such as hydrodynamic dispersion, are not considered in this equation. The dispersion can cause parcels of water to spread transversely to the main direction of groundwater flow and move longitudinally, downgradient, at a faster rate. The dispersion is caused by factors including intermixing of water particles due to differences in interstitial velocity induced by the heterogeneous pore sizes and tortuosity.

3. The migration speeds of most chemicals in a groundwater plume will be retarded by interactions with aquifer materials, especially with clays, organic matter, and metal oxides and hydroxides. This phenomenon will be discussed further in Section 3.5.3.

Example 3.3: Traveling Speed of Leachate through a Compacted Clay Liner

A compacted clay liner (CCL) was installed as the bottom liner of a landfill. The thickness of the CCL is 2 ft, with hydraulic conductivity of $\leq 10^{-7}$ cm/s and effective porosity of 0.25. If the leachate thickness on top of the liner is to be kept ≤ 1 ft, estimate the time needed for leachate to travel through this liner.

Solution:

(a) We need to determine the hydraulic gradient first:

$i = dh/dl$ = (head loss) ÷ (length of the flow path)

= (thickness of the CCL + leachate thickness) ÷ (thickness of the CCL)

= (2 + 1)/(2) = 1.5

Darcy velocity $(v_d) = Ki$

$$= (10^{-7} \text{ cm/s})(1.5) = 1.5 \times 10^{-7} \text{ cm/s}$$

(b) Seepage velocity $(v_s) = v_d/\phi$

$$= (1.5 \times 10^{-7})/(0.25) = 6.0 \times 10^{-7} \text{ cm/s} = 5.2 \times 10^{-2} \text{ cm/day}$$

(c) Time = distance/velocity

$$= (2 \text{ ft})(30.48 \text{ cm/ft}) \div (5.2 \times 10^{-2} \text{ cm/day})$$

$$= 1{,}176 \text{ days} = 3.2 \text{ years}$$

Discussion:

1. The maximum leachate thickness (1 ft) and the maximum hydraulic conductivity of the CCL (10^{-7} cm/s) were used for the worst scenario.

2. Assuming the CCL is intact, it will take 3.2 years for leachate to travel through the 2-ft CCL.

3. The total traveling time will be inversely proportional to the hydraulic gradient and the hydraulic conductivity, but it will be proportional to the thickness of the CCL.

3.2.3 Intrinsic Permeability versus Hydraulic Conductivity

In the soil-venting literature, one may encounter a statement, such as "the soil permeability is 4 darcys"; while in groundwater-remediation literature, one may read about "the hydraulic conductivity is equal to 0.05 cm/s." Both statements describe how permeable the formations are. Are they the same? If not, what is the relationship between the permeability and hydraulic conductivity?

These two terms, *permeability* and *hydraulic conductivity*, are sometimes used interchangeably. However, they do have different meanings. The intrinsic permeability of a porous medium (e.g., subsurface soil or an aquifer) defines its ability to transmit a fluid. It is a property of the medium only and is independent of the properties of the transmitting fluid. That is probably the reason why it is called the "intrinsic" permeability. On the other hand, the hydraulic conductivity of a porous medium depends on the properties of the fluid flowing through it and those of the medium itself.

Hydraulic conductivity is conveniently used to describe the ability of an aquifer to transmit groundwater. A porous medium has a unit hydraulic conductivity if it will transmit a unit volume of groundwater through a unit cross-sectional area (perpendicular to the direction of flow) in a unit time at the prevailing kinematic viscosity and under a unit hydraulic gradient.

The relationship between the intrinsic permeability and hydraulic conductivity is

$$K = \frac{k\rho g}{\mu} \quad \text{or} \quad k = \frac{K\mu}{\rho g} \tag{3.4}$$

where K is the hydraulic conductivity, k is the intrinsic permeability, μ is the fluid viscosity, ρ is the fluid density, and g is the gravitational constant (Note: kinematic viscosity $= \mu/\rho$). The intrinsic permeability has a unit of area as shown in Equation (3.5):

$$k = \frac{K\mu}{\rho g} = \left[\frac{(m/s)(kg/m \cdot s)}{(kg/m^3)(m/s^2)} \right] = [m^2] \tag{3.5}$$

In petroleum industries, the intrinsic permeability of a formation is often expressed in the units of darcy. A formation has an intrinsic permeability of 1 darcy, if it can transmit a flow of 1 cm³/s with a viscosity of 1 centipoise (1 mPa·s) under a pressure gradient of 1 atmosphere/cm acting across an area of 1 cm² (note: 1 Pa = 1 N/m²). That is,

$$1 \text{ darcy} = \frac{(1\,cm^3/s)(10^{-3}\,Pa \cdot s)}{(1\,atm/cm)(1\,cm^2)} \tag{3.6}$$

By substitution of appropriate units for atmosphere (i.e., 1 atm = 1.013 × 10⁵ Pa), it can be shown that

$$1 \text{ darcy} = 0.987 \times 10^{-8}\,cm^2 \tag{3.7}$$

Table 3.2 lists the mass density and viscosity of water under 1 atmosphere. As shown in the table, the density of water from 0°C to 40°C is

TABLE 3.2

Physical Properties of Water under One Atmosphere

Temperature (°C)	Density (g/cm³)	Viscosity (cP)
0	0.999842	1.787
3.98	1.000000	1.567
5	0.999967	1.519
10	0.999703	1.307
15	0.999103	1.139
20	0.998207	1.002
25	0.997048	0.890
30	0.995650	0.798
40	0.992219	0.653

Note: 1 g/cm³ = 1,000 kg/m³ = 62.4 lb/ft³; 1 centipoise (cP) = 0.01 poise = 0.01 g/cm·s = 0.001 Pa·s = 2.1 × 10^{-5} lb·s/ft².

essentially the same, at approximately 1 g/cm³; the viscosity of water decreases with increasing temperature. The viscosity of water at 20°C is 1 centipoise. (Note: This is the viscosity value of the fluid used in defining the darcy unit.)

Example 3.4: Determine Hydraulic Conductivity from a Given Intrinsic Permeability

The intrinsic permeability of a soil core sample is 1 darcy. What is the hydraulic conductivity of this soil for water at 15°C? How about at 25°C?

Solution:

(a) At 15°C,

density of water (15°C) = 0.999703 g/cm³ (from Table 3.2)

viscosity of water (15°C) = 0.01139 poise = 0.01139 g/s·cm (from Table 3.2)

$$K = \frac{k\rho g}{\mu} = \frac{(9.87 \times 10^{-9} \text{cm}^2)(0.999703 \text{ g/cm}^3)(980 \text{ cm/s}^2)}{0.01139 \text{ g/s·cm}} = 8.49 \times 10^{-4} \text{cm/s}$$

$$= (8.49 \times 10^{-4})(2.12 \times 10^4) = 18.0 \text{ gpd/ft}^2 = 18.0 \text{ meinzer}$$

(b) At 25°C,

density of water (25°C) = 0.997048 g/cm³ (from Table 3.2)

viscosity of water (25°C) = 0.00890 poise = 0.00890 g/s·cm (from Table 3.2)

$$K = \frac{k\rho g}{\mu} = \frac{(9.87 \times 10^{-9} \text{ cm}^2)(0.997048 \text{ g/cm}^3)(980 \text{ cm/s}^2)}{0.00890 \text{ g/s} \cdot \text{cm}} = 1.08 \times 10^{-3} \text{cm/s}$$

$$= (1.08 \times 10^{-3})(2.12 \times 10^4) = 23.0 \text{ gpd/ft}^2 = 23.0 \text{ meinzer}$$

Discussion:

1. The value of 2.12×10^4 used in parts (a) and (b) is a conversion factor (1 cm/s = 2.12×10^4 gpd/ft²) from Table 3.1.
2. As mentioned, hydraulic conductivity depends on the properties of the fluid flowing through it. This example illustrates that a porous medium with an intrinsic permeability of 1 darcy has a hydraulic conductivity of 18 gpd/ft² at 15°C and 23 gpd/ft² at 25°C. The hydraulic conductivity of this formation at a higher temperature (25°C) is larger than that at a lower temperature (15°C).
3. The intrinsic permeability is independent of temperature.
4. The unit of gpd/ft² is commonly used by hydrogeologists in the United States. The unit is also named the meinzer after O. E. Meinzer, a pioneering groundwater hydrogeologist with the US Geological Services [1]. The unit of cm/s is more commonly used in soil mechanics. (For example, the hydraulic conductivity of clay liners in landfills is commonly expressed in cm/s.)

From Example 3.4, one can tell that a geologic formation with an intrinsic permeability of one darcy has a hydraulic conductivity of approximately 10^{-3} cm/s or 20 gpd/ft² for transmitting pure water at 20°C. Typical values of intrinsic permeability and hydraulic conductivity for different types of formation are given in Table 3.3.

TABLE 3.3

Typical Values of Intrinsic Permeability and Hydraulic Conductivity

	Intrinsic Permeability	Hydraulic Conductivity	
	(darcy)	(cm/s)	(gpd/ft²)
Clay	10^{-6}–10^{-3}	10^{-9}–10^{-6}	10^{-5}–10^{-2}
Silt	10^{-3}–10^{-1}	10^{-6}–10^{-4}	10^{-2}–1
Silty sand	10^{-2}–1	10^{-5}–10^{-3}	10^{-1}–10
Sand	1–10^2	10^{-3}–10^{-1}	10–10^3
Gravel	10–10^3	10^{-2}–1	10^2–10^4

3.2.4 Transmissivity, Specific Yield, and Storativity

Transmissivity (T) is another commonly used term to describe an aquifer's capacity to transmit water. It represents the amount of water that can be transmitted horizontally by the entire saturated thickness of the aquifer under a hydraulic gradient of one. It is equal to the multiplication product of the aquifer thickness (b) and the hydraulic conductivity (K). Commonly used units for T are m^2/day and gpd/ft.

$$T = Kb \qquad (3.8)$$

An aquifer typically serves two functions: (1) a conduit through which flow occurs and (2) a storage reservoir. This is accomplished by the openings in the aquifer matrix. If a unit of saturated formation is allowed to drain by gravity, not all of the water it contains will be released. The ratio of water that can be drained by gravity to the entire volume of a saturated soil is called *specific yield*, while the part retained is the *specific retention*. Table 3.4 tabulates typical porosity, specific yield, and specific retention of soil, clay, sand, and gravel. The sum of the specific yield and the specific retention of a formation is equal to its porosity.

The specific yield and the specific retention are related to the attraction between water and the formation materials. Clayey formation usually has a lower hydraulic conductivity. This often leads to an incorrect idea that clayey formation has a smaller porosity. As shown in Table 3.4, clay has a much larger porosity than sand and gravel. The porosity of clay can be as high as 50%, but its specific yield is extremely low at 2%. Porosity determines the total volume of water that a formation can store, while specific yield defines the amount that is available to pumping. The low specific yield explains the difficulty of extracting groundwater from clayey aquifers.

When the head in a saturated aquifer changes, water will be taken into or released from storage. Storativity or storage coefficient describes the quantity of water taken into or released from storage per unit change in head

TABLE 3.4

Typical Porosity, Specific Yield, and Specific
Retention of Selected Materials

	Porosity (%)	Specific Yield (%)	Specific Retention (%)
Soil	55	40	15
Clay	50	2	48
Sands	25	22	3
Gravel	20	19	1

Source: [2].

per unit area. It is a dimensionless quantity. The response of a confined aquifer to the change of water head is different from that of an unconfined aquifer. When the head declines, a confined aquifer remains saturated; the water is released from storage due to the expansion of water and the compaction of aquifer. The amount of release is exceedingly small. On the other hand, the water table rises or falls with change of head in an unconfined aquifer. As the water level changes, water drains from or enters into the pore spaces. This storage or release is mainly due to the specific yield. It is also a dimensionless quantity. For unconfined aquifers, the storativity is practically equal to the specific yield, and ranges typically between 0.1 and 0.3. The storativity of confined aquifers is substantially smaller and generally ranges between 0.0001 and 0.00001, and that for leaky confined aquifers is in the range of 0.001. A small storativity implies that it will require a larger pressure change (or gradient) to extract groundwater at a specific flow rate [2].

The volume of groundwater (V) drained from an aquifer can be determined from Equation (3.9):

$$V = S \times A \times \Delta h \qquad (3.9)$$

where S is the storativity, A is the area of the aquifer, and Δh is the change in head.

Example 3.5: Estimate Loss of Storage in Aquifers due to Change in Head

An unconfined aquifer has an area of 5 square miles. The storativity of this aquifer is 0.15. The water table fell 0.8 ft during a recent drought. Please estimate the amount of water lost from storage.

If the aquifer is confined and its storativity is 0.0005, what would be the amount lost for a decrease of 0.8 ft in head?

Solution:

(a) Inserting the values into Equation (3.9), we obtain the volume of water drained from the unconfined aquifer:

$V = (0.15)[(5)(5280)^2 \text{ ft}^2](0.8 \text{ ft}) = 1.67 \times 10^7 \text{ ft}^3 = 1.25 \times 10^8 \text{ gal}$

(b) For the confined aquifer:

$V = (0.0005)[(5)(5280)^2 \text{ ft}^2](0.8 \text{ ft}) = 5.58 \times 10^4 \text{ ft}^3 = 4.17 \times 10^5 \text{ gal}$

Discussion:

For the same amount of change in head, the water lost in the unconfined aquifer is 300 times more, which is the ratio of the two storativity values (0.15/0.0005 = 300).

3.2.5 Determine Groundwater Flow Gradient and Flow Direction

Having a good knowledge of the gradient and direction of groundwater flow is vital to groundwater remediation. The gradient and direction of flow have great impacts on selection of remediation schemes to control plume migration, such as location of the pumping wells and groundwater extraction rates, etc.

Estimates of the gradient and direction of groundwater flow can be made from a minimum of three groundwater elevations. The general procedure is described here, and an example follows.

Step 1: Locate the three surveyed points on a map to scale.

Step 2: Connect the three points and mark their water-table elevations on the map.

Step 3: Subdivide each side of the triangle into a number of segments of equal size. (Each segment represents an increment of groundwater elevation.)

Step 4: Connect the points of equal values of elevation (equipotential lines), which then form the groundwater contours.

Step 5: Draw a line that passes through and is perpendicular to each equipotential line. This line marks direction of flow.

Step 6: Calculate the groundwater gradient from the formula, $i = dh/dl$.

Example 3.6: Estimate the Gradient and Direction of Groundwater Flow from Three Groundwater Elevations

Three groundwater monitoring wells were installed at an impacted site. Groundwater elevations were determined from a recent survey of these wells, and the values were marked on a map. Estimate the flow gradient and direction of the groundwater flow in the underlying aquifer.

Solution:

(a) Water elevations (36.2′, 35.6′, and 35.4′) were measured at three monitoring wells and marked on the map.

(b) These three points are connected by straight lines to form a triangle.

(c) Subdivide each side of the triangle into a number of segments of equal intervals. For example, subdivide the line connecting point A (36.2′) and point B (35.6′) into three intervals. Each interval represents a 0.2′-increment in elevation.

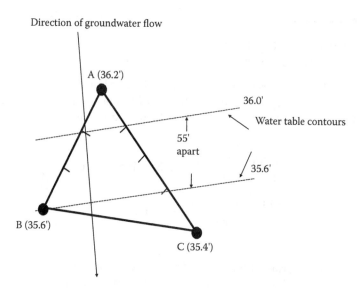

FIGURE 3.1
Determination of groundwater gradient and direction.

(d) Connect the points of equal values of elevation (equipotential lines), which then form the groundwater contours. Here, we connect the elevations of 35.6′ and 36.0′ to form two contour lines.

(e) Draw a line that passes through and is perpendicular to each equipotential line and mark it as the groundwater flow direction.

(f) Measure the distance between two contour lines, 55 ft in this example.

(g) Calculate the groundwater gradient from the formula, $i = dh/dl$:

$$i = (36.0 - 35.6)/(55) = 0.0073$$

See Figure 3.1.

Discussion:

The groundwater elevations, especially those of the water-table aquifers, may change with time. Consequently, the groundwater flow gradient and direction would change. Periodic surveys of the groundwater elevation may be necessary if fluctuation of the water table is suspected. Off-site pumping, seasonal change, and recharge are some of the reasons that may cause the fluctuation of the water-table elevation.

3.3 Groundwater Pumping
3.3.1 Steady-State Flow in a Confined Aquifer

Equation (3.10) describes steady-state flow of a confined aquifer (an artesian aquifer) from a fully penetrating well. A fully penetrating well means that the groundwater can enter at any level from the top to the bottom of the aquifer.

$$Q = \frac{Kb(h_2 - h_1)}{528 \log(r_2/r_1)} \quad \text{(American Practical Units)}$$

$$= \frac{2.73 \, Kb(h_2 - h_1)}{\log(r_2/r_1)} \quad \text{(SI Units)} \tag{3.10}$$

where
Q = pumping rate or well yield (in gpm, or m³/day)
h_1, h_2 = static head measured from the aquifer bottom (in ft or m)
r_1, r_2 = radial distance from the pumping well (in ft or m)
b = thickness of the aquifer (in ft or m)
K = hydraulic conductivity of the aquifer (in gpd/ft² or m/day)

Many assumptions were made to derive this equation. Several references and other groundwater hydrology books provide more detailed treatment of this subject [1, 3–5].

Hydraulic conductivity is often determined from aquifer tests (see Section 3.4 for details). Equation (3.10) can be readily modified to estimate the hydraulic conductivity of a confined aquifer if steady-state drawdown, flow rate, and aquifer thickness data are available.

$$K = \frac{528 \, Q \log(r_2/r_1)}{b(h_2 - h_1)} \quad \text{(American Practical Units)}$$

$$= \frac{Q \log(r_2/r_1)}{2.73b(h_2 - h_1)} \quad \text{(SI Units)} \tag{3.11}$$

Another parameter, specific capacity, can also be used to estimate the hydraulic conductivity of an aquifer. Let us define the specific capacity as

$$\text{Specific capacity} = \frac{Q}{s_w} \tag{3.12}$$

where
Q = the well discharge rate (extraction rate), in gpm
s_w = drawdown in the pumping well, in ft

For example, if a well produces 50 gpm and the drawdown in the well is 5 ft, the specific capacity of this pumping well is 10 gpm/ft (i.e., it will produce 10 gpm for each foot of available drawdown). A rough estimate on transmissivity (in gpd/ft) can be obtained by multiplying the specific yield (in gpm/ft) by 2,000 for confined aquifers and 1,550 for unconfined aquifers [2]. The hydraulic conductivity (in gpd/ft^2) can then be determined by dividing the transmissivity with the aquifer thickness (in ft).

Example 3.7: Steady-State Drawdown from Pumping a Confined Aquifer

A confined aquifer (30 ft or 9.1 m in thickness) has a piezometric surface 80 ft (24.4 m) above the bottom confining layer. Groundwater is being extracted from a 4-in. (0.1 m)-diameter fully penetrating well.

The pumping rate is 40 gpm (0.15 m^3/min). The aquifer is relatively sandy, with a hydraulic conductivity of 200 gpd/ft^2. Steady-state drawdown of 5 ft (1.5 m) is observed in a monitoring well 10 ft (3.0 m) from the pumping well. Estimate

- The drawdown 30 ft (9.1 m) away from the well
- The drawdown in the pumping well.

Solution:

(a) First let us determine h_1 (at $r_1 = 10$ ft):

$$h_1 = 80 - 5 = 75 \text{ ft} \quad (\text{or} = 24.4 - 1.5 = 22.9 \text{ m})$$

Use Equation (3.10):

$$40 = \frac{(200)(30)(h_2 - 75)}{528 \log(30/10)} \quad \Rightarrow h_2 = 76.7 \text{ ft}$$

or

$$(0.15)(1,440) = \frac{2.73\,[(200)(0.0410)](9.1)(h_2 - 22.9)}{\log(9.1/3.0)} \quad \Rightarrow h_2 = 23.4 \text{ m}$$

So drawdown at 30 ft (9.1 m) away $= 80 - 76.7 = 3.3$ ft (or $= 24.4 - 23.4 = 1.0$ m)

(b) To determine the drawdown at the pumping well, set r at the wellbore = well radius = (2/12) ft:

$$40 = \frac{(200)(30)(h_2 - 75)}{528 \log[(2/12)/10]} \quad \Rightarrow h_2 = 68.7 \text{ ft}$$

So, drawdown in the extraction well $= 80 - 68.7 = 11.3$ ft

Discussion:

1. In part (a), 0.041 is the conversion factor to convert the hydraulic conductivity from gpd/ft² to m/day. The factor was taken from Table 3.1.

2. Calculations in part (a) have demonstrated that the results would be the same by using two different systems of units.

3. The $(h_1 - h_2)$ term can be replaced by $(s_2 - s_1)$, where s_1 and s_2 are the drawdown at r_1 and r_2, respectively.

4. The same equation can also be used to determine the radius of influence, where drawdown is equal to zero. This topic is discussed further in Chapter 6.

Example 3.8: Estimate Hydraulic Conductivity of a Confined Aquifer from Steady-State Drawdown Data

Use the following information to estimate the hydraulic conductivity of a confined aquifer:

- Aquifer thickness = 30.0 ft (9.1 m)
- Well diameter = 4 in. (0.1 m)
- Well perforation = fully penetrating
- Groundwater extraction rate = 20 gpm
- Steady-state drawdown
  - = 2.0 ft observed in a monitoring well 5 ft from the pumping well
  - = 1.2 ft observed in a monitoring well 20 ft from the pumping well

Solution:

Inserting the data into Equation (3.11), we obtain:

$$K = \frac{528\,Q\log(r_2/r_1)}{b(h_2 - h_1)} = \frac{(528)(20)\log(20/5)}{(30)(2.0-1.2)} = 397 \text{ gpd/ft}^2$$

Discussion:

The $(h_1 - h_2)$ term can be replaced by $(s_2 - s_1)$, where s_1 and s_2 are the drawdown at r_1 and r_2, respectively.

Example 3.9: Estimate Hydraulic Conductivity of a Confined Aquifer Using Specific Capacity

Use the drawdown data of the pumping well in Example 3.7 to estimate the hydraulic conductivity of the aquifer:

- Aquifer thickness = 30 ft
- Pumping rate = 40 gpm
- Steady-state drawdown in the well = 11.3 ft

Solution:

(a) First let us determine the specific capacity. Use Equation (3.12):

$$\text{Specific capacity} = \frac{Q}{s_w} = \frac{40}{11.3} = 3.54 \text{ gpm/ft}$$

(b) The transmissivity of the aquifer can be estimated as:

$$T = (3.54)(2,000) = 7,080 \text{ gpd/ft}$$

(c) The hydraulic conductivity of the aquifer can be estimated as:

$$K = T/b = 7,080/30 = 236 \text{ gpd/ft}^2$$

Discussion:

The calculated hydraulic conductivity (236 gpd/ft²) from this exercise is not much different from the value specified in Example 3.7 (200 gpd/ft²).

3.3.2 Steady-State Flow in an Unconfined Aquifer

The equation describing steady-state flow of an unconfined aquifer (water-table aquifer) from a fully penetrating well may be written as follows:

$$Q = \frac{K(h_2^2 - h_1^2)}{1,055 \log(r_2/r_1)} \quad \text{(American Practical Units)}$$

$$= \frac{1.366\, K(h_2^2 - h_1^2)}{\log(r_2/r_1)} \quad \text{(SI Units)} \quad (3.13)$$

All the terms are as defined for Equation (3.10).

Equation (3.13) can be easily modified to calculate the hydraulic conductivity of an unconfined aquifer if data of two steady-state drawdowns and flow rate are available.

$$K = \frac{1,055\, Q \log(r_2/r_1)}{(h_2^2 - h_1^2)} \quad \text{(American Practical Units)}$$

$$= \frac{Q \log(r_2/r_1)}{1.366\, (h_2^2 - h_1^2)} \quad \text{(SI Units)} \quad (3.14)$$

The specific capacity, defined by Equation (3.12), can also be used to estimate the hydraulic conductivity of an unconfined aquifer.

Example 3.10: Steady-State Drawdown from Pumping an Unconfined Aquifer

A water-table aquifer is 80 ft (24.4 m) thick. Groundwater is being extracted from a 4-in. (0.1 m)-diameter fully penetrating well.

The pumping rate is 40 gpm (0.15 m³/min). The aquifer is relatively sandy, with a hydraulic conductivity of 200 gpd/ft². Steady-state drawdown of 5 ft (1.5 m) is observed in a monitoring well 10 ft (3.0 m) from the pumping well. Estimate:

- The drawdown 30 ft (9.1 m) away from the well
- The drawdown in the pumping well

Solution:

(a) First let us determine h_1 (at $r_1 = 10$ ft):

$$h_1 = 80 - 5 = 75 \text{ ft (or} = 24.4 - 1.5 = 22.9 \text{ m)}$$

Use Equation (3.14):

$$40 = \frac{(200)(h_2{}^2 - 75^2)}{1,055 \log(30/10)} \Rightarrow h_2 = 75.7 \text{ ft}$$

or

$$(0.15)(1440) = \frac{1.366 \, [(200)(0.0410)](h_2{}^2 - 22.9^2)}{\log(9.1/3.0)} \Rightarrow h_2 = 23.1 \text{ m}$$

So drawdown at 30 ft (9.1 m) away $= 80 - 75.7 = 4.3$ ft (or $= 24.4 - 23.1 = 1.3$ m)

(b) To determine the drawdown at the pumping well, set r at the well = well radius = (2/12) ft

$$40 = \frac{(200)(h_2{}^2 - 75^2)}{1,055 \log[(2/12)/10]} \Rightarrow h_2 = 72.5 \text{ ft}$$

So, drawdown in the extraction well $= 80 - 72.5 = 7.5$ ft

Discussion:

1. In the equation for confined aquifers, the $(h_1 - h_2)$ term can be replaced by $(s_2 - s_1)$, where s_1 and s_2 are the drawdown at r_1 and r_2, respectively. However, no analogy can be made here, that is, $(h_2{}^2 - h_1{}^2)$ cannot be replaced by $(s_1{}^2 - s_2{}^2)$.

2. The same equation can also be used to determine the radius of influence, where drawdown is equal to zero. More discussions on this topic are given in Chapter 6.

Example 3.11: Estimate Hydraulic Conductivity of an Unconfined Aquifer from Steady-State Drawdown Data

Use the following information to estimate the hydraulic conductivity of an unconfined aquifer:

- Aquifer thickness = 30.0 ft (9.1 m)
- Well diameter = 4 in. (0.1 m)
- Well perforation = fully penetrating
- Groundwater extraction rate = 20 gpm
- Steady-state drawdown = 2.0 ft observed in a monitoring well 5 ft from the pumping well
  = 1.2 ft observed in a monitoring well 20 ft from the pumping well

Solution:

First we need to determine h_1 and h_2:

$h_1 = 30.0 - 2.0 = 28.0$ ft
$h_2 = 30.0 - 1.2 = 28.8$ ft

Inserting the data into Equation (3.14), we obtain:

$$K = \frac{(1,055)(20)\log(20/5)}{(28.8^2 - 28^2)} = 280 \text{ gpd/ft}^2$$

Discussion:

Drawdown and flow rate data in Examples 3.8 and 3.11 (one for a confined aquifer and the other for an unconfined aquifer) are the same; however, the calculated hydraulic conductivity values are different. In these examples, the hydraulic conductivity of the unconfined aquifer is smaller, but it delivers the same flow rate with the same drawdown because the unconfined aquifer has a larger storage coefficient. Refer to Section 3.2.4 for the discussion on the storage coefficient.

Example 3.12: Estimate Hydraulic Conductivity of an Unconfined Aquifer Using Specific Capacity

Use the pumping and drawdown data in Example 3.10 to estimate the hydraulic conductivity of the aquifer:

- Aquifer thickness = 80 ft
- Pumping rate = 40 gpm
- Steady-state drawdown in the well = 7.5 ft

Solution:

(a) First let us determine the specific capacity. Use Equation (3.12),

$$\text{Specific capacity} = \frac{Q}{s_w} = \frac{40}{7.5} = 5.3 \text{ gpm/ft}^2$$

(b) The transmissivity of the aquifer can be estimated as:

$$T = (5.3)(1,550) = 8,220 \text{ gpd/ft}$$

(c) The hydraulic conductivity of the aquifer can be estimated as:

$$K = T/b = 8,220/80 = 103 \text{ gpd/ft}^2$$

Discussion:

The calculated hydraulic conductivity (103 gpd/ft²) from this exercise has the same order of magnitude as the value specified in Example 3.10 (200 gpd/ft²).

3.4 Aquifer Tests

In Section 3.3, methods using the steady-state drawdown data (Equations 3.11 and 3.14) were described to estimate the hydraulic conductivity of aquifers. For a groundwater-remediation project, it is often required to have a good estimate of the hydraulic conductivity before the full-scale groundwater extraction. Grain-size analysis of aquifer materials and bench-scale testing on core samples can provide some limited information. For more accurate estimates, aquifer tests are often conducted.

Pumping tests and slug tests are two common types of aquifer tests. In a typical pumping test, groundwater is extracted from a pumping well at a constant rate. (Other pumping schemes are also feasible, but not as popular.) The time-dependent drawdowns (or recovery) in a pumping well and in a few monitoring wells are recorded. The data are then analyzed to determine the hydraulic conductivity and storativity. The pumping test is recommended because it provides information on subsurface hydrogeology over a large area (the area affected by the pumping) and gives a realistic estimate of the pumping rate for the full-scale groundwater extraction. Many remediation systems have been incorrectly designed and installed for a flow rate much larger than the extraction wells could yield due to lack of accurate aquifer information. In addition, analysis of groundwater extracted during a pump test will give engineers a more realistic estimation of the COC concentrations for treatment system design than those just based on the data from sampling of monitoring wells. The main disadvantage of a pumping test is the expenses associated with conductance of the test, data analysis, and treatment and disposal of the extracted water.

A cheaper alternative to a pumping test is a slug test in which a slug of known volume is inserted into the water inside a well. The rate at which water level falls is collected and analyzed. The disadvantages of a slug test are (1) it provides the hydrological information related only to the vicinity of the well, and (2) it provides no information for estimates of the extracted COC concentrations once the full-scale remediation program starts. No further discussion on slug tests will be given here.

The flow in the aquifer during a pumping test is considered to be under unsteady-state conditions. Three common methods are used to analyze the unsteady-state data: (1) Theis curve matching, (2) the Cooper–Jacob straight-line method, and (3) the distance–drawdown method.

3.4.1 Theis Method

The drawdown for confined aquifers under unsteady-state pumping was first solved by C. V. Theis as:

$$s = \frac{114.6Q}{T}\left[-0.5772 - \ln(u) + u - \frac{u^2}{2\cdot 2!} + \frac{u^3}{3\cdot 3!} - \frac{u^4}{4\cdot 4!} + \ldots\right] \quad \text{(American Practical Units)}$$

$$= \frac{Q}{4\pi T}\left[-0.5772 - \ln(u) + u - \frac{u^2}{2\cdot 2!} + \frac{u^3}{3\cdot 3!} - \frac{u^4}{4\cdot 4!} + \ldots\right] \quad \text{(SI Units)} \quad (3.15)$$

where the argument u is dimensionless and given as

$$u = \frac{1.87r^2 S}{Tt} \quad \text{(American Practical Units)}$$

$$= \frac{r^2 S}{4Tt} \quad \text{(SI Units)} \quad (3.16)$$

where
s = drawdown at time t (in ft or m)
Q = constant pumping rate (in gpm or m³/day)
r = radial distance from the pumping well to the observation well (in ft or m)
S = aquifer storativity (dimensionless)

T = aquifer transmissivity (in gpd/ft or m²/day)
t = time since pumping started (in days)

The infinite-series term in Equation (3.15) (the terms inside the square bracket) is often called the well function and designated as $W(u)$. Tabulated values of $W(u)$ as a function of u can be found in groundwater hydrology books. (The well function tables have become obsolete because of the convenience of hand calculators and personal computers.) A type-curve approach is often developed to match the time and drawdown data to the curve of $W(u)$ versus $1/u$. From the match points, the transmissivity and storativity can be determined. There are computer programs commercially available for Theis curve matching. This subsection will provide one example of using the Theis equation, but no examples for the curve matching will be given.

Example 3.13: Estimate Unsteady-State Drawdown of a Confined Aquifer Using the Theis Equation

A pumping well is installed in a confined aquifer. Use the following information to estimate the drawdown at a distance 20 ft away from the well after one day of pumping:

- Aquifer thickness = 30.0 ft
- Groundwater extraction rate = 20 gpm
- Aquifer hydraulic conductivity = 400 gpd/ft²
- Aquifer storativity = 0.005

Solution:

(a) $T = Kb = (400)(30) = 12,000$ gpd/ft

(b) Inserting the data into Equation (3.16), we obtain

$$u = \frac{1.87 r^2 S}{Tt} = \frac{1.87(20 \text{ ft})^2 (0.005)}{(12,000 \text{ gpd/ft})(1 \text{ day})} = 3.12 \times 10^{-4}$$

(c) Substitute the value of u in the well function to obtain its value:

$$W(u) = \left[-0.5772 - \ln(3.12 \times 10^{-4}) + (3.12 \times 10^{-4}) - \frac{(3.12 \times 10^{-4})^2}{2 \cdot 2!} \right.$$
$$\left. + \frac{(3.12 \times 10^{-4})^3}{3 \cdot 3!} - \frac{(3.12 \times 10^{-4})^4}{4 \cdot 4!} + \ldots \right]$$

$$= 7.50$$

(d)   The drawdown can then be determined from Equation (3.15):

$$s = (114.6)(20)(7.50)/(12{,}000) = 1.43 \text{ ft}$$

Discussion:

For small u values, the third and later terms in the well function can be truncated without causing a significant error.

3.4.2 Cooper–Jacob's Straight-Line Method

As shown in Example 3.13, the higher terms in the well function become negligible for small u values. Cooper and Jacob (1946) [6] pointed out that, for small u values, the Theis equation can be modified to the following form without significant errors:

$$s = \frac{264Q}{T} \log\left[\frac{0.3Tt}{r^2S}\right] \text{ (American Practical Units)}$$

$$= \frac{0.183Q}{T} \log\left[\frac{2.25Tt}{r^2S}\right] \text{ (SI Units)} \qquad (3.17)$$

where the symbols represent the same terms as in Equation (3.15).

As shown in Equation 3.16, the value of u becomes small as t increases and r decreases. So Equation (3.17) is valid after sufficient pumping time and at a short distance from the well ($u < 0.05$). It can be seen from Equation (3.17) that, at any specific location (r = constant), s varies linearly with $\log[(\text{constant})t]$. The Cooper–Jacob straight-line method is to plot drawdown vs. pumping-time data from a pumping test on semilog paper; most of the data should fall on a straight line. From the plot, the slope, Δs (the change in drawdown per one log cycle of time), and the intercept, t_0, of the straight line at zero drawdown can be derived. The following relationships can then be used to determine the transmissivity and storativity of the aquifer:

$$T = \frac{264Q}{\Delta s} \text{ (American Practical Units)} \quad = \frac{0.183Q}{\Delta s} \text{ (SI Units)} \qquad (3.18)$$

$$S = \frac{0.3T t_0}{r^2} \text{ (American Practical Units)} = \frac{2.25T t_0}{r^2} \text{ (SI Units)} \qquad (3.19)$$

where Δs is in ft or m, t_0 in days, and the other symbols represent the same terms as in Equation (3.15).

Example 3.14: Analysis of Pumping Test Data Using Cooper–Jacob's Straight-Line Method

A pumping test (Q = 120 gpm) was conducted on a confined aquifer (aquifer thickness = 30.0 ft). The time-drawdown data at a distance 150 ft away from the well were collected and shown in the following table.

Use the Cooper–Jacob's straight-line method to determine the hydraulic conductivity and storativity of the aquifer.

Time Since Pumping Started, T (min)	Drawdown, s (ft)
7	0.15
20	0.45
80	0.90
200	1.16

Solution:

(a) The data are first plotted on a semilog scale, as seen in Figure 3.2. From the plot, we find Δs = 0.7 ft.

(b) Use Equation (3.18):

$$T = \frac{264Q}{\Delta s} = \frac{(264)(50)}{0.7} = 18{,}860 \text{ gpd/ft}$$

(c) Hydraulic conductivity can then be found as
$$K = T/b = (18{,}860)/(30) = 629 \text{ gpd/ft}^2$$

(d) From the plot, we find the intercept, t_0 = 4.5 min = 3.1×10^{-3} day

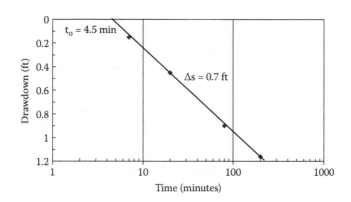

FIGURE 3.2
Cooper–Jacob's straight-line method for pumping data analysis.

Use Equation (3.19) to find the storativity:

$$S = \frac{0.3Tt_0}{r^2} = \frac{(0.3)(18,860)(0.0031)}{(150)^2} = 0.00078$$

Discussion:

1. At $t = 7$ min (0.00486 day) and $r = 150$ ft, u is equal to

$$u = \frac{1.87r^2S}{Tt} = \frac{1.87(150 \text{ ft})^2(0.00078)}{(18,860 \text{ gpd/ft})(0.00486 \text{ day})} = 0.36$$

2. At $t = 50$ min, u will be smaller than 0.05.

3.4.3 Distance–Drawdown Method

It can be seen from Equation (3.17) that, at any specific time (t = constant), s varies linearly with $\log[(\text{constant})/r^2]$. Based on this relationship and simultaneous drawdown measurements in at least three observation wells, each at a different distance from the pumping well, a semilog distance–drawdown graph can be constructed. From the plot, the slope, Δs (the change in drawdown per one log cycle of distance), and the intercept, r_0, of the straight line at zero drawdown can be derived. The following relationships can then be used to determine the transmissivity and storativity of the aquifer:

$$T = \frac{528Q}{\Delta s} \text{ (American Practical Units)} = \frac{0.366Q}{\Delta s} \text{ (SI Units)} \qquad (3.20)$$

$$S = \frac{0.3Tt}{r_0^2} \text{ (American Practical Units)} = \frac{2.25Tt}{r_0^2} \text{ (SI Units)} \qquad (3.21)$$

where Δs is in ft or m, r_0 is in ft or m, and the other symbols represent the same terms as in Equation (3.15).

The three methods described here for analysis of pumping test data are mainly for confined aquifers. Extraction groundwater from an unconfined aquifer is more complicated. The extracted water comes from two mechanisms: (1) water from the elastic storage due to the decline in pressure, as in the case of the confined aquifer, and (2) water from drainage of the declining water table. There would be three distinct phases of time-drawdown relations in unconfined aquifers. As time progresses, the rate of drawdown decreases and flow becomes essentially horizontal (when the effects of gravity drainage become much smaller). The time-drawdown data can then be analyzed using the three methods described previously [1]. A more practical approach is to ensure that the duration of the pumping test exceeds the

TABLE 3.5
Suggested Guidelines for Pumping Tests of Unconfined Aquifer

Predominant Aquifer Material	Minimum Pumping Time (hours)
Medium sand and coarser materials	4
Fine sand	30
Silt and clay	170

Source: [5].

suggested guidelines in Table 3.5 [5]. As shown in the table, the suggested pumping duration increases with the tightness of aquifer. A minimum of 7 days pumping is suggested for silty or clayey aquifers.

Example 3.15: Analysis of Pumping Test Data Using the Distance–Drawdown Method

A pumping test ($Q = 120$ gpm) was conducted on a confined aquifer (aquifer thickness = 30.0 ft). The distance–drawdown data (at $t = 90$ min) were collected from three monitoring wells and shown in the following table.

Use the distance–drawdown method to determine the hydraulic conductivity and storativity of the aquifer.

Distance from the Pumping Well (ft)	Drawdown, s (ft)
50	1.55
150	0.90
300	0.50

Solution:

(a) The data are first plotted on a semilog scale. (See Figure 3.3) From the plot, we find $\Delta s = 1.4$ ft.

(b) Use Equation (3.20) to find the transmissivity:
$$T = \frac{528Q}{\Delta s} = \frac{(528)(50)}{1.4} = 18,860 \text{ gpd/ft}$$

(c) Hydraulic conductivity can then be found as:
$K = T/b = (18,860)/(30) = 629$ gpd/ft^2

(d) From the plot, we find the intercept, $r_0 = 650$ ft

(e) Use Equation (3.21) and $t = 80$ min $= 0.0555$ day to find the storativity:
$$S = \frac{0.3Tt}{r_0^2} = \frac{(0.3)(18,860)(0.0555)}{(650)^2} = 0.00074$$

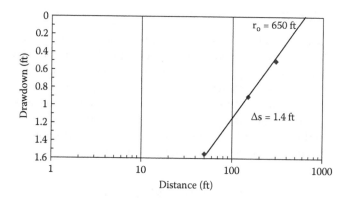

FIGURE 3.3
Distance–drawdown method for pumping data analysis.

Discussion:

1. As expected, the slope of the straight line in the distance–drawdown is twice that in Cooper–Jacob's straight-line plot (for the same hydraulic conductivity and pumping rate).

2. At $t = 90$ min (0.0625 day) and $r = 300$ ft, u is equal to

$$u = \frac{1.87r^2S}{Tt} = \frac{1.87(300 \text{ ft})^2(0.00074)}{(18,860 \text{ gpd/ft})(0.0625 \text{ day})} = 0.11$$

3. At $r < 206$ ft, u will be smaller than 0.05.

3.5 Migration Velocity of the Dissolved Plume

As VOC spills enter into the subsurface, the materials may move downward as free product or be dissolved into the infiltrating water and then move downward by gravity. This liquid may travel deep enough to get in contact with the underlying aquifer and form a dissolved plume in the aquifer. This section will discuss the migration of the dissolved plume, which is relatively simpler than the transport of the COCs in the vadose zone. This discussion is applicable not only to VOCs, but to other types of COCs, such as heavy metals. Transport in the vadose zone is discussed in Section 3.6.

3.5.1 Advection–Dispersion Equation

Design and selection of optimal remediation schemes, such as the number and locations of extraction wells, often require prediction of the COC distribution in the subsurface over time. These predictions are then used to evaluate different remediation scenarios. To make such predictions, we need to couple the equation describing the flow with the equation of mass balance. More discussions on the mass balance concept can be found in Chapter 4.

To describe the fate and transport of a COC, the one-dimensional form of the advection–dispersion equation can be expressed as:

$$\frac{\partial C}{\partial t} = D\frac{\partial^2 C}{\partial x^2} - v\frac{\partial C}{\partial x} \pm \text{RXNs} \qquad (3.22)$$

where C is the COC concentration, D is the dispersion coefficient, v is the velocity of the fluid flow, t is the time, and RXNs represents the reactions. Equation (3.22) is a general equation, and it is applicable to describe the fate and transport of COCs in the vadose zone or in the groundwater. The first term of Equation (3.22) describes the change in COC concentration in fluid, contained within a specific volume of an aquifer or a vadose zone, with time. The first term on the right-hand side describes the net dispersive flux of the COC in and out of the fluid in this volume, and the second term describes the net advective flux. The last term represents the amount of COC that may be added or lost to the fluid in this volume due to physical, chemical, and/or biological reactions. For plume migration in groundwater, v is the groundwater velocity that can be determined from Darcy's law and the porosity of the aquifer (i.e., Equation 3.3).

3.5.2 Diffusivity and Dispersion Coefficient

The dispersion term in Equation (3.22) accounts for both the molecular diffusion and hydraulic dispersion. The molecular diffusion, strictly speaking, is due to concentration gradient (i.e., the concentration difference). The COC tends to diffuse away from the higher concentration zone, and this can occur even when the fluid is not moving. The hydraulic dispersion here is mainly caused by flow in porous media. It results from (1) velocity variation within a pore, (2) different pore geometrics, (3) divergence of flow lines around the soil grains in the porous media, and (4) the aquifer heterogeneity [7].

The unit of the dispersion coefficient is $(\text{length})^2/(\text{time})$. Field studies of the dispersion coefficient revealed that it varies with groundwater velocity. They show that the dispersion coefficient is relatively constant at low velocities (where the molecular dispersion dominates), but increases linearly with velocity as the groundwater velocity increases (when the hydraulic dispersion dominates). The dispersion coefficient can be written as the sum of two terms: effective molecular diffusion coefficient, D_d, and hydraulic dispersion coefficient, D_h:

$$D = D_d + D_h \tag{3.23}$$

The effective molecular diffusion coefficient can be obtained from molecular diffusion coefficient, D_0, as

$$D_d = (\xi)(D_0) \tag{3.24}$$

where ξ is the tortuosity factor that accounts for the increased distance that COCs need to travel to get around the soil grains. Typical ξ values are in the range of 0.6 to 0.7 [7].

The hydraulic dispersion coefficient is proportional to the groundwater flow velocity as

$$D_h = (\alpha)(v) \tag{3.25}$$

where α is the dispersivity. The hydraulic dispersion coefficient is scale dependent; its value has been observed to increase with increasing transport distance. The longitudinal dispersivity values from field tracer tests and model calibration of COC plumes are found to be in the range of 10 to 100 m, which is much higher than that from column studies in the laboratories.

The molecular diffusion coefficients of COCs in dilute aqueous solutions are very much smaller than in gases at atmospheric pressure, usually ranging from 0.5 to 2×10^{-5} cm²/s at 25°C (compared with typical values of 0.05 to 0.5 cm²/s in the gaseous phase, as shown in Table 2.5). Values of molecular diffusion coefficients of selected compounds are shown in Table 3.6.

TABLE 3.6

Values of Diffusion Coefficients of Selected Compounds in Water

Compound	Temperature (°C)	Diffusion Coefficient (cm²/s)
Acetone	25	1.28×10^{-5}
Acetonitrile	15	1.26×10^{-5}
Benzene	20	1.02×10^{-5}
Benzoic acid	25	1.00×10^{-5}
Butanol	15	0.77×10^{-5}
Ethylene glycol	25	1.16×10^{-5}
Propanol	15	0.87×10^{-5}

Source: [8].

The diffusion coefficient of a chemical in water can be estimated by using the Wilke–Chang method [8]:

$$D_0 = \frac{5.06 \times 10^{-7} T}{\mu_w V^{0.6}} \tag{3.26}$$

where
D_0 = diffusion coefficient, in cm^2/s
T = temperature, in K
μ_w = viscosity of water, in cP (see Table 3.2)
V = molal volume of the solute at its normal boiling point, in cm^3/g mole

The molal volume can be derived using the LeBas method and data in Table 3.7 [8].

TABLE 3.7

Additive-Volume Increments for Calculation of Molal Volumes

	Increment (cm^3/g mole)
Carbon	14.8
Hydrogen	3.7
Oxygen (except as noted below)	7.4
in methyl esters and ethers	9.1
in ethyl esters and ethers	9.9
in higher esters and ethers	11.0
in acids	12.0
joined to S, P, N	8.3
Nitrogen	
doubly bonded	15.6
in primary amines	10.5
in secondary amines	12.0
Bromine	27
Chlorine	24.6
Ring	
three-membered	−6.0
four-membered	−8.5
five-membered	−11.5
six-membered	−15.0
naphthalene	−30.0
anthracite	−47.5

Source: [8].

The diffusion coefficient of a compound can be estimated using the diffusion coefficient of another compound of similar species, their molecular weights, and the following relationship:

$$\frac{D_1}{D_2} = \sqrt{\frac{MW_2}{MW_1}} \qquad (3.27)$$

As shown in Equation (3.27), the diffusion coefficient is inversely proportional to the square root of its molecular weight. The heavier the COC, the harder it is for it to diffuse through the fluid. Temperature also has an influence on the diffusion coefficient. From Equation (3.26), we can see the diffusion coefficient in water is proportional to the temperature and inversely proportional to the fluid viscosity. The water viscosity (μ_w) decreases with increasing temperature and, consequently, the diffusion coefficient increases with temperature and the following relationship applies:

$$\frac{D_0 @ T_1}{D_0 @ T_2} = \left(\frac{T_1}{T_2}\right)\left(\frac{\mu_w @ T_2}{\mu_w @ T_1}\right) \qquad (3.28)$$

Example 3.16: Estimate the Diffusion Coefficient Using the LeBas Method

Estimate the diffusion coefficient of toluene in a dilute aqueous solution at 20°C using the LeBas method.

Solution:

(a) The formula of toluene is $C_6H_5CH_3$. It consists of a benzene ring (six carbon member) and a methyl group.

Viscosity of water at 25°C = 0.89 cP (from Table 3.2)

$T = 275 + 20 = 293K$

Molal volume is determined from the sum of the volume increments (Table 3.7)

$C = (14.8)(7) = 103.6$

$H = (3.7)(8) = 29.6$

Six-membered ring = –15.0

So, $V = 103.6 + 29.6 - 15.0 = 118.2$ cm³/g mole

(b) Use Equation (3.26) to find the diffusion coefficient:

$$D_0 = \frac{5.06 \times 10^{-7}(293)}{(0.89)(118.2)^{0.6}} = 0.95 \times 10^{-5} \text{ cm}^2/\text{s}$$

Example 3.17: Estimate the Diffusion Coefficient at Different Temperatures

The diffusion coefficient of benzene in a dilute aqueous solution at 20°C is 1.02×10^{-5} cm²/s (Table 3.6). Use this reported value to estimate:

- The diffusion coefficient of toluene in a dilute aqueous solution at 20°C
- The diffusion coefficient of benzene in a dilute aqueous solution at 25°C

Solution:

(a) The MW of toluene ($C_6H_5CH_3$) is 92, and the MW of benzene (C_6H_6) is 78.

Use Equation (3.27):

$$\frac{D_1}{D_2} = \frac{(1.02 \times 10^{-5})}{D_2} = \sqrt{\frac{92}{78}}$$

So the diffusion coefficient of toluene at 20°C = 0.94×10^{-5} cm²/s

(b) Viscosity of water at 20°C = 1.002 cP (from Table 3.2)
Viscosity of water at 25°C = 0.89 cP (from Table 3.2)
Use Equation (3.28):

$$\frac{(1.02 \times 10^{-5})}{D_0 @ 298K} = \left(\frac{293}{298}\right)\left(\frac{0.89}{1.002}\right)$$

So the diffusion coefficient of benzene at 25°C = 1.17×10^{-5} cm²/s

Discussion:

1. The diffusion coefficient of toluene estimated from that of benzene is 0.94×10^{-5} cm²/s, which is essentially the same as that from the LeBas method, 0.94×10^{-5} cm²/s (Example 3.16).
2. The diffusion coefficient of benzene at 25°C is about 15% larger than that at 20°C.

Example 3.18: Relative Importance of Molecular Diffusion and Hydraulic Dispersion

Benzene from USTs at a site leaked into the underlying aquifer. The hydraulic conductivity of the aquifer is 500 gpd/ft² and it has an effective porosity of 0.4. The groundwater temperature is 20°C. The dispersivity is found to be 2 m. Estimate the relative importance between the hydraulic dispersion and the molecular diffusion for the dispersion of the benzene plume in the following two cases:

1. The hydraulic gradient = 0.01
2. The hydraulic gradient = 0.0005

Solution:

(a) The hydraulic conductivity of the aquifer = 500 gpd/ft²

$$= (500)(4.73 \times 10^{-5}) = 0.024 \text{ cm/s (Use the conversion factor in Table 3.1.)}$$

Use Equations (3.1) and (3.2) to find out the groundwater velocity (for gradient = 0.01)

$$v_s = \frac{(0.024)(0.01)}{0.4} = 6 \times 10^{-4} \text{ cm/s}$$

The molecular diffusion coefficient of benzene (at 20°C) = 1.02×10^{-5} cm²/s (Table 3.6).

From Equation (3.23), $D = D_d + D_h$

The effective molecular diffusion coefficient can be obtained as (Equation 3.24):

$$D_d = \xi(D_0) = (0.65)(1.02 \times 10^{-5}) = 0.66 \times 10^{-5} \text{ cm}^2/\text{s}$$

The hydraulic dispersion coefficient can be determined as (Equation 3.25):

$$D_h = \alpha(v) = (200 \text{ cm})(6 \times 10^{-4} \text{ cm/s}) = 12{,}000 \times 10^{-5} \text{ cm}^2/\text{s}$$

The hydraulic dispersion coefficient is much larger than the diffusion coefficient. Therefore, the hydraulic dispersion will be the dominant mechanism for the dispersion of COCs.

(b) For a smaller gradient, the groundwater will move more slowly, and the dispersion coefficient will be proportionally smaller. The effective molecular diffusion coefficient will be the same as 0.66×10^{-5} cm²/s.

Use Equations (3.1) and (3.2) to find the groundwater velocity (for gradient = 0.0005):

$$v_s = \frac{(0.024)(0.0005)}{0.4} = 3.0 \times 10^{-5} \text{ cm/s}$$

The hydraulic dispersion coefficient can then be determined as (Equation 3.25):

$$D_h = \alpha(v) = (200 \text{ cm})(3.0 \times 10^{-5} \text{ cm/s}) = 600 \times 10^{-5} \text{ cm}^2/\text{s}$$

The hydraulic dispersion coefficient is still much larger than the diffusion coefficient at this relatively flat gradient of 0.0005.

Discussion:

In the second case, the groundwater movement is very slow at 3.0×10^{-5} cm/s (or 31 ft/yr), but the hydraulic dispersion is still the dominant mechanism (for dispersivity = 2 m). The diffusion coefficient will become more important only if the flow rate and/or the dispersivity is smaller. Nonetheless, the molecular diffusion accounts for a common phenomenon that the plume usually extends slightly upstream of the entry point into the aquifer.

3.5.3 Retardation Factor for Migration in Groundwater

Physical, chemical, and biological processes in subsurface that can affect the fate and transport of COCs include biotic degradation, abiotic degradation, dissolution, ionization, volatilization, and adsorption. Adsorption of COCs is probably the most important and most studied mechanism for removal of COCs from the dissolved plume in groundwater. If adsorption is the primary removal mechanism in the subsurface, the reaction term in Equation (3.22) can then be written as $(\rho_b/\phi)\partial S/\partial t$, where ρ_b is the dry bulk density of soil (or the aquifer matrix), ϕ is the porosity, t is time, and S is the COC concentration adsorbed onto the aquifer solids.

When the COC concentration is low, a linear adsorption isotherm is usually valid. (See Section 2.4.3 for further discussion on the adsorption isotherms.) Assume a linear adsorption isotherm (e.g., $S = K_p C$), thus

$$\frac{\partial S}{\partial C} = K_p \tag{3.29}$$

The following relationship can then be derived:

$$\frac{\partial S}{\partial C} = \left(\frac{\partial S}{\partial C}\right)\left(\frac{\partial C}{\partial t}\right) = K_p \frac{\partial C}{\partial t} \tag{3.30}$$

Substitute Equation (3.30) into Equation (3.22) and rearrange the equation

$$\frac{\partial C}{\partial t} + \left(\frac{\rho_b}{\phi}\right)K_p \frac{\partial C}{\partial t} = \left(1 + \frac{\rho_b K_p}{\phi}\right)\frac{\partial C}{\partial t} = D\frac{\partial^2 C}{\partial x^2} - v\frac{\partial C}{\partial x} \tag{3.31}$$

By dividing both sides by $(1 + \rho_b K_p/\phi)$, Equation (3.31) can be simplified into the following form:

$$\frac{\partial C}{\partial t} = \frac{D}{R}\frac{\partial^2 C}{\partial x^2} - \frac{v}{R}\frac{\partial C}{\partial x} \tag{3.32}$$

where

$$R = 1 + \frac{\rho_b K_p}{\phi} \tag{3.33}$$

The parameter, R, is often called the retardation factor (dimensionless) and has a value ≥ 1. Equation (3.32) is essentially the same as Equation (3.22), except that the reaction term in Equation (3.22) is taken care of by R (Equation 3.33). The retardation factor reduces the impact of dispersion and migration velocity by a factor of R. All of the mathematical solutions that are used to solve the transport of inert tracers can be used for the transport of the COCs if the groundwater velocity and the dispersion coefficient are divided by the retardation factor. From the definition of R, we can tell that R is a function of ρ_b, ϕ, and K_p. For a given aquifer, ρ_b and ϕ would be the same for different COCs. Consequently, the larger the partition coefficient, the larger is the retardation factor.

Example 3.19: Determination of the Retardation Factor

The aquifer underneath a site is impacted by several organic compounds, including benzene, 1,2-dichloroethane (DCA), and pyrene. Estimate their retardation factors using the following data from the site assessment:

- Effective aquifer porosity = 0.40
- Dry bulk density of the aquifer materials = 1.6 g/cm³
- Fraction of organic carbon of the aquifer materials = 0.015
- $K_{oc} = 0.63 K_{ow}$

Solution:

(a) From Table 2.5,

$\mathrm{Log}(K_{ow}) = 2.13$ for benzene $\rightarrow K_{ow} = 135$
$\mathrm{Log}(K_{ow}) = 1.53$ for 1,2-DCA $\rightarrow K_{ow} = 34$
$\mathrm{Log}(K_{ow}) = 4.88$ for pyrene $\rightarrow K_{ow} = 75,900$

(b) Using the given relationship, $K_{oc} = 0.63 K_{ow}$, we obtain:

$K_{oc} = (0.63)(135) = 85$ (benzene)
$K_{oc} = (0.63)(34) = 22$ (1,2-DCA)
$K_{oc} = (0.63)(75,900) = 47,800$ (pyrene)

(c) Using Equation (2.26), $K_p = f_{oc} K_{oc}$, and $f_{oc} = 0.015$, we obtain:

$K_p = (0.015)(85) = 1.275$ (benzene)
$K_p = (0.015)(22) = 0.32$ (1,2-DCA)
$K_p = (0.015)(47,800) = 717$ (pyrene)

(d) Use Equation (3.33) to find the retardation factor:

$$R = 1 + \frac{\rho_b K_p}{\phi} = 1 + \frac{(1.6)(1.275)}{0.4} = 6.10 \quad \text{for benzene}$$

$$R = 1 + \frac{\rho_b K_p}{\phi} = 1 + \frac{(1.6)(0.32)}{0.4} = 2.28 \quad \text{for 1,2-DCA}$$

$$R = 1 + \frac{\rho_b K_p}{\phi} = 1 + \frac{(1.6)(717)}{0.4} = 2,869 \quad \text{for pyrene}$$

Discussion:

Pyrene is very hydrophobic with a large K_p value, and its retardation factor is much larger than those of benzene and 1,2-DCA.

3.5.4 Migration of Dissolved Plume

The retardation factor relates the plume migration velocity to the groundwater seepage velocity as

$$R = \frac{V_s}{V_p} \quad \text{or} \quad V_p = \frac{V_s}{R} \tag{3.34}$$

where V_s is the groundwater seepage velocity and V_p is the velocity of the dissolved plume. When the value of R is equal to unity (for inert compounds), the compound will move at the same speed as the groundwater flow without any "retardation." When $R = 2$, for example, the COC will move at half of the groundwater flow velocity.

Example 3.20: Migration Speed of Dissolved Plume in Groundwater

The aquifer underneath a site is impacted by several organic compounds including benzene, 1,2-dichloroethane (DCA), and pyrene. A recent groundwater monitoring in September 2013 indicated that 1,2-DCA and benzene have traveled 250 m and 20 m downgradient, respectively, while no pyrene compounds were detected in the downgradient wells.

Estimate the time when the leachates first entered the aquifer. The following data were obtained from the site assessment:

- Effective aquifer porosity = 0.40
- Aquifer hydraulic conductivity = 30 m/day

- Groundwater gradient = 0.005
- Dry bulk density of aquifer materials = 1.6 g/cm³
- Fraction of organic carbon of the aquifer materials = 0.015
- $K_{oc} = 0.63\ K_{ow}$

Briefly discuss your results and list possible factors that may cause your estimate to differ from the true value.

Solution:

(a) Use Equation (3.1) to find the Darcy velocity:
$$v = k_i = (30)(0.005) = 0.15 \text{ m/d}$$

(b) Use Equation (3.3) to find the groundwater velocity (i.e., the seepage velocity, or the interstitial velocity):
$$v_s = v/\phi = (0.15)/(0.4) = 0.375 \text{ m/d}$$

(c) Use Equation (3.34) and the values of R from Example 3.19 to determine the migration speeds of the plumes:
$$v_p = (0.375)/(6.10) = 0.061 \text{ m/d} = 22.4 \text{ m/yr (benzene)}$$
$$v_p = (0.375)/(2.28) = 0.164 \text{ m/d} = 60.0 \text{ m/yr (1,2-DCA)}$$
$$v_p = (0.375)/(2,864) = 0.000131 \text{ m/d} = 0.048 \text{ m/yr (pyrene)}$$

(d) The time for 1,2-DCA to travel 250 meters can be found as:
$$t = (\text{distance})/(\text{migration speed})$$
$$= (250 \text{ m})/(60.0 \text{ m/yr}) = 4.17 \text{ yr} = 4 \text{ years and 2 months}$$
So 1,2-DCA entered the aquifer in July of 2009.

(e) The time for benzene to travel 50 meters can be found as:
$$t = (50 \text{ m})/(22.4 \text{ m/yr}) = 2.23 \text{ yr} = 2 \text{ years and 3 months}$$
So benzene entered the aquifer in June of 2011.

Discussion:

1. The estimates are the times when benzene and 1,2-DCA first entered the aquifer. The data given are insufficient to estimate the time the leachates traveled through the vadose zone and, consequently, the time the leaks started from the sources (e.g., leaky USTs).

2. The retardation factor of 1,2-DCA is smaller; therefore, its migration speed in the vadose zone would be faster. This helps to explain the fact that 1,2-DCA entered the aquifer earlier than benzene.

3. The migration speed of pyrene is extremely slow, 0.042 m/yr; therefore, it was not detected in the downstream monitoring wells. Most, if not all, of the pyrene compounds will be adsorbed onto the soil in the vadose zone. The pyrene may travel in the aquifer by adsorbing onto the colloidal particles.

4. The estimates are crude because many factors may affect the accuracy of the estimates. Factors include uncertainty in the values of the hydraulic conductivity, porosity, groundwater gradient, K_{ow}, f_{oc}, etc. Neighborhood pumping will affect the groundwater gradient and, consequently, the migration of the plume. Other subsurface reactions such as oxidation and biodegradation may also have large impacts on the fate and transport of these COCs.

Example 3.21: Migration Speed of the Dissolved Plume in Groundwater

Results of a recent quarterly groundwater monitoring (July of 2013) at a site indicate that the edge of the dissolved TCE plume has advanced 200 m in the past 5 years. The groundwater gradient was determined to be 0.01 from this round of monitoring. Using a value of 4.0 for the retardation factor and effective aquifer porosity of 0.35, what would be your estimate of the aquifer hydraulic conductivity? Also, because of the drought, an adjacent facility (downgradient from the site) pumped out a great amount of groundwater in 2010. How will this affect your estimate?

Solution:

(a) The migration speeds of the plume, v_p:

$$= (\text{distance})/(\text{time}) = (200)/5 = 40 \text{ m/yr}$$

(b) Use Equation (3.34) and the value of R to find the groundwater velocity, v_s:

$$v_p = v_s/R = 40 = v_s/4 \rightarrow v_s = 160 \text{ m/yr}$$

(c) Use Equation (3.3) to find the Darcy velocity, v_d:

$$v_s = v_d/\phi = 160 = (v_d)/(0.35) \rightarrow v_d = 56 \text{ m/yr}$$

(d) Use Equation (3.1) to find the aquifer hydraulic conductivity:

$$v_d = Ki = (K)(0.01) = 56 \text{ m/yr} \rightarrow K = 5,600 \text{ m/yr} = 15.3 \text{ m/day}$$

Discussion:

The neighborhood pumping during the drought would increase the natural groundwater gradient. During the pumping period, the groundwater moved faster, and so did the plume. In other words, the plume would have traveled a shorter distance without the pumping. The hydraulic conductivity of the aquifer would be smaller than this estimate, 15.3 m/day.

Example 3.22 Migration Speed of the Dissolved
Plume and Partition of COCs

The toluene concentration of the groundwater in an aquifer was determined to be 500 ppb. Assuming no free product phase present, estimate the partition of toluene in the two phases, i.e., dissolved in liquid and adsorbed onto the aquifer materials.

From the RI work, the following parameters were determined:

- Retardation factor = 4.0
- Porosity = 0.35
- Dry bulk density of the aquifer matrix = 1.6 g/cm^3.

Strategy:

To determine the partition between the liquid phase and solid phase, we need to know the partition coefficient. The partition coefficient can be found from the retardation factor.

Solution:

(a) Use Equation (3.33) to determine the partition coefficient, K_p:

$$R = 1 + \frac{\rho_b K_p}{\phi} = 4 = 1 + \frac{(1.6)K_p}{0.35}$$

So $K_p = 0.656$ L/kg

Use Equation (2.24) to find the toluene concentration on the aquifer solid, S

$$S = K_p C = (0.656)(0.5) = 0.328 \text{ mg/kg}$$

(b) Basis: 1 L of aquifer

Mass dissolved in liquid = $(V)(\phi)(C)$
$$= (1 \text{ L})(0.35)(0.5 \text{ mg/L}) = 0.175 \text{ mg}$$
Mass adsorbed on the aquifer solids = $(V)(\rho_b)(S)$
$$= (1 \text{ L})(1.6 \text{ kg/L})(0.328 \text{ mg/kg})$$
$$= 0.525 \text{ mg}$$
$$\% \text{ mass in liquid} = (0.175) \div [(0.175) + (0.525)]$$
$$= 25\% \text{ mass in liquid}$$

Discussion:

This example illustrates that the majority of toluene was attached to the aquifer solids; only 25% was in the dissolved phase. This partially explains why the cleanup takes a long time for groundwater remediation using the pump-and-treat method.

3.6 COC Transport in the Vadose Zone

The travel of COCs in the vadose zone can occur in three ways: (1) volatilizing into the air void and traveling as vapor, (2) becoming dissolved into the soil moisture and/or into the infiltrating water and then traveling with the liquid, and (3) moving downward by gravity as the immiscible phase. This section describes these transport pathways.

3.6.1 Liquid Movement in the Vadose Zone

Liquid flow through the vadose zone can be described by a differential equation, and its one-dimensional form is

$$\frac{\partial}{\partial z}\left[K\frac{\partial \Psi}{\partial z}\right] + \frac{\partial K}{\partial z} = \frac{\partial \theta_w}{\partial \Psi}\frac{\partial \Psi}{\partial t} \tag{3.35}$$

where K is the hydraulic conductivity, θ_w is the volumetric water content, ψ is the soil water pressure head (the sum of the gravity potential and the moisture potential), z is the distance, and t is the time. The major differences between this equation and the equation for one-dimensional groundwater flow (i.e., Darcy's law) are: (1) the hydraulic conductivity in the vadose zone is a function of ψ, and hence of θ_w, and (2) the pressure head is a function of time. These make Equation (3.35) nonlinear, time-dependent, and more difficult to solve than the simple Darcy's equation. (If K is a constant and pressure head is independent of time, then Equation (3.35) can be simplified to the Darcy's equation.)

The hydraulic conductivity of a vadose zone is the largest at water saturation and decreases as the water content decreases. As the moisture content decreases, air occupies most of the pore void and leaves a smaller cross-sectional area for water transport. Consequently, the hydraulic conductivity decreases. At a very low moisture content, the water film covering the soil particles becomes very thin. The attractive forces between the water molecules and the soil grains become so strong that no water will move. At this point, the hydraulic conductivity is approaching zero. The hydraulic conductivity at a given moisture can be found from the relative permeability for that moisture, k_r (a dimensionless term), and the hydraulic conductivity at saturation, K_s, as

$$K = k_r K_s \tag{3.36}$$

The relative hydraulic conductivity varies from 1.0 at 100% saturation to 0.0 at 0% saturation.

The travel of the dissolved COC in the vadose zone can be described by an advection–dispersion equation, and its one-dimensional form is

$$\frac{\partial(\theta_w C)}{\partial t} = \frac{\partial^2(\theta_w DC)}{\partial z^2} - \frac{\partial(\theta_w vC)}{\partial z} \pm \text{RXNs} \tag{3.37}$$

This equation is similar to the one for the saturated zone (i.e., Equation 3.22), except the volumetric water content, θ_w, is a variable, and the velocity and dispersion coefficient depend on the moisture content. The dispersion coefficient is analogous to the dispersion term in the saturated zone, except v is a function of the moisture content, as

$$D = D_d + D_h = \xi \times D_0 + \alpha \times v(\theta_w) \tag{3.38}$$

Example 3.23: Estimate the Hydraulic Conductivity in the Vadose Zone

A subsurface soil is relatively sandy and has a hydraulic conductivity of 500 gpd/ft^2 when the soil is saturated. Estimate its hydraulic conductivity (a) when the water saturation is 40% and (b) when the water saturation is 90%. The relative permeability for sand at 40% saturation is 0.02, and that at 90% saturation is 0.44.

Solution:

(a) Use Equation (3.36) to find the hydraulic conductivity at 40% saturation:

$K = (0.02)(500) = 10$ gpd/ft^2

(b) Use Equation (3.36) to find the hydraulic conductivity at 90% saturation:

$K = (0.44)(500) = 220$ gpd/ft^2

Discussion:

The water saturation is the percentage of the pore space that is occupied by the water: 100% for saturated soil and 0% for dry soil. At 40% water saturation, the hydraulic conductivity is close to zero, and at 90% water saturation, the hydraulic conductivity is 44% of the maximum value.

3.6.2 Gaseous Diffusion in the Vadose Zone

Under nonpumping conditions, the molecular diffusion is the prime mechanism for gas-phase transport. The transport equation can be expressed by Fick's law, and its one-dimensional form is

$$\xi_a \phi_a D \frac{\partial^2 G}{\partial x^2} = \frac{\partial(\phi_a G)}{\partial t} \tag{3.39}$$

where D is the free-air diffusion coefficient, G is the COC concentration in the gas phase, ϕ_a is the air-filled porosity, and ξ_a is the air-phase tortuosity factor. The ξ_a term accounts for the diffusion taking place within a porous medium rather than in an open air space. It can be estimated from empirical equations such as the Millington-Quirk equation [9]:

$$\xi_a = \frac{\phi_a^{10/3}}{\phi_t^2} \tag{3.40}$$

where ϕ_t is the total porosity, which is the sum of the air-filled porosity and the volumetric water content ($\phi_t = \phi_a + \phi_w$). The air-phase tortuosity factor varies from zero, when the entire pore space is occupied by water (saturated condition), to about 0.8, when the porosity is high and the medium is dry.

The values of free-air diffusion coefficients for selected compounds can be found in Table 2.5. The free-air diffusion coefficient is generally 10,000 times higher than that in a dilute aqueous solution. The diffusion coefficient can also be estimated from the diffusion coefficient of another compound of similar species and their molecular weights by the following relationship (same as that for liquid in Equation 3.27):

$$\frac{D_1}{D_2} = \sqrt{\frac{MW_2}{MW_1}} \tag{3.41}$$

The diffusion coefficient is inversely proportional to the square root of its molecular weight. The heavier the compound, the harder it is for it to diffuse through the air. Temperature can have an influence on the diffusion coefficient. The diffusion coefficient increases with temperature, and the following relationship applies:

$$\frac{D_0 @ T_1}{D_0 @ T_2} = \left(\frac{T_1}{T_2}\right)^m \tag{3.42}$$

where T is the temperature in Kelvin. Theoretically, the exponent, m, should be 1.5; however, experimental data indicate that it ranges from 1.75 to 2.0.

Example 3.24: Estimate the Air-Phase Tortuosity Factor

A subsurface soil is relatively sandy and has a porosity of 0.45. Estimate its air-phase tortuosity factor:

1. When the volumetric water content is 0.3
2. When the volumetric water content is 0.05.

Solution:

(a)   For $\phi_w = 0.3$ and $\phi_t = 0.45$,

$$\phi_a = 0.45 - 0.3 = 0.15$$

Use Equation (3.40) to find the air-phase tortuosity factor at $\phi_w = 0.3$:

$$\varsigma_a = \frac{(0.15)^{10/3}}{(0.45)^2} = 0.0088$$

(b)   For $\phi_w = 0.05$ and $\phi_t = 0.45$,

$$\phi_a = 0.45 - 0.05 = 0.40$$

Use Equation (3.40) to find the air-phase tortuosity factor at $\phi_w = 0.05$

$$\varsigma_a = \frac{(0.40)^{10/3}}{(0.45)^2} = 0.23$$

Discussion:

The volumetric water content here is the percentage of total soil volume (not the void volume) occupied by water. For this case, the air-phase tortuosity becomes approximately 25 times larger when the volumetric water content drops from 0.3 to 0.05.

Example 3.25:   Estimate the Diffusion Coefficient at Different Temperatures

The diffusion coefficient of benzene in dilute aqueous solution at 20°C is 1.02 $\times$ 10^{-5} cm^2/s (Table 3.6) and the free-air diffusion coefficient of benzene is 0.092 cm^2/s at 25°C (Table 2.5). Use these reported values to estimate:

1. The ratio of the diffusion coefficients of benzene in free air and in a dilute aqueous solution at 20°C
2. The free air diffusion coefficient of toluene at 20°C.

Solution:

(a)   Use Equation (3.42) and $m = 2$ (assumed) to determine the free air diffusion coefficient of benzene at 20°C:

$$\frac{0.092}{D_0 @ T_2} = \left(\frac{298}{293}\right)^2$$

So the free air diffusion coefficient of benzene at 20°C = 0.089 cm^2/s.

The ratio between the free air and liquid diffusion coefficients
$$= (0.089) \div (1.02 \times 10^{-5}) = 8{,}720$$

(b) The MW of toluene ($C_6H_5CH_3$) is 92, and the MW of benzene (C_6H_6) is 78.

Use Equation (3.41) to determine the diffusion coefficient:

$$\frac{D_1}{D_2} = \frac{(0.089)}{D_2} = \sqrt{\frac{92}{78}}$$

So the diffusion coefficient of toluene at 20°C = 0.082 cm²/s.

Discussion:

1. The diffusion coefficient of benzene in free air is 8,720 higher than that in the dilute aqueous phase.

2. The diffusion coefficient of toluene estimated from that of benzene and the molecular weight relationship (0.082 cm²/s) is essentially the same as that in Table 2.5 (0.083 cm²/s).

3.6.3 Retardation Factor for COC Vapor Migration in the Vadose Zone

For an air stream flowing through a porous medium, the gas-phase retardation factor can be derived as [9]

$$R_a = 1 + \frac{\rho_b K_p}{\phi_a H} + \frac{\phi_w}{\phi_a H} \tag{3.43}$$

where ρ_b is the dry bulk density, K_p is the soil–water partition coefficient, H is the Henry's constant, ϕ_a is the air-filled porosity, and ϕ_w is the volumetric water content.

This retardation factor will be a constant if ϕ_w does not change. It is analogous to the retardation factor, R, for the movement of COCs in an aquifer. The movement of the COC in the void of the vadose zone will be retarded by a factor of R_a. The second term on the right-hand side of Equation (3.43) represents the partitioning of the COCs between the vapor phase, the soil moisture phase, and the solid phase. The third term represents the partitioning between the vapor phase and the solid phase. As the COC in the vapor phase moves through the air-filled pores, the migration rate of the COC in the air is slower than that of the air itself, because of the loss of its mass to the soil moisture and to the soil organic carbon.

Under the condition of no advective flow, the gas-phase retardation factor can be defined as the ratio of the diffusion rate of an inert compound such as nitrogen to the diffusion rate of the COC. Under advective flow, it can be used as the relative measure to compare the migration rates of compounds with different retardation factors. For soil-venting application, the air-phase retardation factor is also the minimum number of pore volumes that must pass through the impacted zone to clean it up. It is considered as the minimum because this approach ignores the effects of mass-transfer limitations among the phases, subsurface heterogeneity, and unequal travel time from the outer edge of the plume to the vapor-extraction well [9].

As shown in Equation (3.43), the air-phase retardation factor increases with ϕ_w and K_p, but decreases with Henry's constant. Higher moisture content means a larger water reservoir to retain the COCs, and a larger K_p value indicates that soil has a larger organic content or the COC is more hydrophobic. On the other hand, a compound with a larger Henry's constant would have a stronger tendency to volatilize into the air void. The Henry's constant increases with increasing temperature and, thus, a smaller air-phase retardation factor at a higher temperature. Therefore, for a soil-venting application, at higher temperatures, fewer pore volumes of air need to be moved through the impacted zone to remove the COCs.

Example 3.26: Determination of the Air-Phase Retardation Factor

The vadose zone underneath a site is impacted by several organic compounds, including benzene, 1,2-dichloroethane (DCA), and pyrene.

Estimate the air-phase retardation factor using the following data from the site assessment:

- Vadose zone soil porosity = 0.40
- Volumetric water content = 0.15
- Dry bulk density of soil = 1.6 g/cm³
- Fraction of organic carbon of soil = 0.015
- Temperature of the formation = 25°C
- $K_{oc} = 0.63 K_{ow}$

Solution:

(a) From Table 2.5,

$H = 5.55$ atm/M for benzene (at 25°C)

Use Table 2.4 to convert it to a dimensionless value:

$H^* = H/RT = (5.55)/[(0.082)(298)] = 0.227$ (for benzene)

Similarly for 1,2-DCA (the Henry's constant value in the table is for 20°C, so we use this value for 25°C as an approximate value) and pyrene:

$H^* = H/RT = (0.98)/[(0.082)(298)] = 0.04$ (for 1,2-DCA)

$H^* = H/RT = (0.005)/[(0.082)(298)] = 0.0002$ (for pyrene)

(b) From Example 3.19,

$K_p = (0.015)(85) = 1.275$ (for benzene)

$K_p = (0.015)(22) = 0.32$ (for 1,2-DCA)

$K_p = (0.015)(47,800) = 717$ (for pyrene)

(c) Use Equation (3.43) to find the air-phase retardation factor:

$$R_a = 1 + \frac{\rho_b K_p}{\phi_a H} + \frac{\phi_w}{\phi_a H} = 1 + \frac{(1.6)(1.275)}{(0.25)(0.227)} + \frac{(0.15)}{(0.25)(0.227)} = 39.6 \quad \text{(benzene)}$$

$$R_a = 1 + \frac{\rho_b K_p}{\phi_a H} + \frac{\phi_w}{\phi_a H} = 1 + \frac{(1.6)(0.32)}{(0.25)(0.04)} + \frac{(0.15)}{(0.25)(0.04)} = 67.2 \quad \text{(1,2-DCA)}$$

$$R_a = 1 + \frac{\rho_b K_p}{\phi_a H} + \frac{\phi_w}{\phi_a H} = 1 + \frac{(1.6)(717)}{(0.25)(0.0002)} + \frac{(0.15)}{(0.25)(0.0002)} = 2.3 \times 10^7 \quad \text{(pyrene)}$$

Discussion:

Pyrene is very hydrophobic and has a low Henry's constant. Its air-phase retardation factor is much larger than those of benzene and 1,2-DCA.

References

1. Fetter Jr., C.W. 1980. *Applied hydrogeology*. Columbus, OH: Charles E. Merrill.
2. USEPA. 1990. Ground water, Vol. I: Ground water and contamination. EPA/625/6-90/016a. Washington, DC: US Environmental Protection Agency.
3. Driscoll, F.G. 1986. *Groundwater and wells*. 2nd. ed. St. Paul, MN: Johnson Division.
4. Freeze, R.A., and R.A. Cherry. 1979. *Groundwater*. Englewood Cliffs, NJ: Prentice Hall.
5. Todd, D.K. 1980. *Groundwater hydrology*. 2nd. ed. New York: John Wiley & Sons.
6. Cooper, H.H., and C.E. Jacob. 1946. A generalized graphical method for evaluating formation constants and summarizing well-field history. *Am Geophys. Union Trans.* 27 (4), 526–534.
7. USEPA. 1989. Transport and face of contaminants in the subsurface. EPA/625/4-89/019. Washington, DC: US Environmental Protection Agency.
8. Sherwood, T.K., R.L. Pigford, and C.R. Wilke. 1975. *Mass transfer under reduced gravity*. New York: McGraw-Hill.
9. USEPA. 1991. Site characterization for subsurface remediation. EPA/625/4-91/026. Washington, DC: US Environmental Protection Agency.

4

Mass-Balance Concept and Reactor Design

4.1 Introduction

Various treatment processes are employed in groundwater and soil remediation. Treatment processes are generally classified into physical, chemical, biological, and thermal processes. A treatment system often consists of a series of unit operations/processes, which form a process train. Each unit operation/process contains one or more reactors. A reactor can be considered as a vessel in which the processes occur. Environmental engineers are often in charge of, or at least participate in, conceptual and preliminary design of the treatment system. The conceptual and preliminary design typically includes selection of treatment processes, determination of reactor types, and sizing of the reactors.

For system design, treatment processes should be chosen first by screening the alternatives. Many factors should be considered in selection of treatment processes. Common selection criteria are implementability, effectiveness, cost, and regulatory consideration. In other words, an optimal process would be the one that is implementable, effective in removal of compounds of concern (COCs), cost efficient, and in compliance with the regulatory requirements.

Once the treatment processes are selected for a remediation project, engineers will then design the reactors. Preliminary reactor design usually includes selecting appropriate reactor types, sizing reactors, and determining the number of reactors needed and their optimal configuration. To size the reactors, engineers first need to know if the desirable reactions or activities would occur in the reactors and what the optimal operating conditions, such as temperature and pressure, would be. Information from chemical thermodynamics and, more practically, a bench- and/or pilot-scale study would provide answers to these questions. If the desired reactions are feasible, the engineers then need to determine the rates of these reactions, which is a subject of chemical kinetics. The reactor size is then determined, based on mass loading to the reactor, reaction rate, type of the reactor, and target effluent levels.

This chapter introduces the mass-balance concept, which is the basis for process design. Then it presents reaction kinetics, as well as types, configuration, and sizing of reactors. This chapter covers the topics on how to determine the rate constant, removal efficiency, optimal arrangement of reactors, required residence time, and reactor size for specific applications.

4.2 Mass-Balance Concept

The mass-balance (or material balance) concept serves as a basis for designing environmental engineering systems (reactors). The mass-balance concept is nothing but conservation of mass. Matter can neither be created nor destroyed (a nuclear process is one of the few exceptions), but it can be changed in form. The fundamental approach is to show the changes occurring in the reactor by the mass-balance analysis. The following is a general form of a mass-balance equation:

$$\begin{bmatrix} \text{Rate of mass} \\ \text{ACCUMULATED} \end{bmatrix} = \begin{bmatrix} \text{Rate of mass} \\ \text{IN} \end{bmatrix} - \begin{bmatrix} \text{Rate of mass} \\ \text{OUT} \end{bmatrix} \pm \begin{bmatrix} \text{Rate of mass} \\ \text{GENERATED or} \\ \text{DESTROYED} \end{bmatrix}$$

$$(4.1)$$

Performing a mass balance on an environmental engineering system is just like balancing a checking account. The rate of mass accumulated (or depleted) in a reactor can be viewed as the rate that money is accumulated in (or depleted from) the checking account. How fast the balance changes depends on how much and how often the money is deposited and withdrawn (rate of mass input and output), interest incurred (rate of mass generated), and bank charges for service and ATM fees (rate of mass destroyed).

In using the mass-balance concept to analyze an environmental engineering system, we usually begin by drawing a process flow diagram and employing the following procedure:

Step 1: Draw system boundaries or boxes around the unit processes/ operations or flow junctions to facilitate calculations.

Step 2: Place known flow rates and concentrations of all streams, sizes and types of reactors, as well as operating conditions such as temperature and pressure on the diagram.

Step 3: Calculate and convert all known mass inputs, outputs, and accumulation/disappearance to the same units and place them on the diagram.

Step 4: Mark unknown (or the ones to be found) inputs, outputs, and accumulation/disappearance on the diagram.

Step 5: Perform the necessary analyses/calculations using the procedures described in this chapter.

A few special cases or reasonable assumptions can simplify the general mass-balance equation, Equation (4.1), and make the analysis easier. Three common ones are presented here:

No reactions occurring: If the system has no chemical reactions occurring, there will be no increases or decreases of compound mass due to reactions. The mass-balance equation would become:

$$\begin{bmatrix} \text{Rate of mass} \\ \text{ACCUMULATED} \end{bmatrix} = \begin{bmatrix} \text{Rate of mass} \\ \text{IN} \end{bmatrix} - \begin{bmatrix} \text{Rate of mass} \\ \text{OUT} \end{bmatrix} \quad (4.2)$$

Batch reactor: For a batch reactor, there is no input into and output out of the reactor. The mass-balance equation can be simplified into:

$$\begin{bmatrix} \text{Rate of mass} \\ \text{ACCUMULATED} \end{bmatrix} = \pm \begin{bmatrix} \text{Rate of mass} \\ \text{GENERATED or} \\ \text{DESTROYED} \end{bmatrix} \quad (4.3)$$

Steady-state conditions: To maintain the stability of treatment processes, treatment systems are usually operated under steady-state conditions after a start-up period. A steady-state condition basically means that flow and concentrations at any locations within the treatment process train are not changing with time. Although the concentration and/or flow rate of the influent waste stream entering a soil/groundwater system typically fluctuate, engineers may want to incorporate devices such as equalization tanks to dampen the fluctuation. This is especially true for treatment processes that are sensitive to fluctuations in mass loading (e.g., biological processes).

For a reactor under a steady-state condition, although reactions are occurring, the rate of mass accumulation in the reactor would be zero. Consequently, the left-hand side term of Equation (4.1) becomes zero. The mass-balance equation can then be reduced to:

$$0 = \begin{bmatrix} \text{Rate of mass} \\ \text{IN} \end{bmatrix} - \begin{bmatrix} \text{Rate of mass} \\ \text{OUT} \end{bmatrix} \pm \begin{bmatrix} \text{Rate of mass} \\ \text{GENERATED or} \\ \text{DESTROYED} \end{bmatrix} \quad (4.4)$$

Assumption of steady-state is frequently used in the analysis of flow reactors. It should be noted that a batch reactor is operated under unsteady state because the concentration in the reactor is changing, and it is not a flow reactor because there is no flow in and out of the reactor when it is in operation. The general mass-balance equation (i.e., Equation 4.1) can also be expressed as:

$$V\frac{dC}{dt} = \sum Q_{in}C_{in} - \sum Q_{out}C_{out} \pm (V \times \gamma) \qquad (4.5)$$

where V is the volume of the system (reactor), C is the concentration, Q is the flow rate, and γ is the reaction rate. The following sections will demonstrate the role of the reaction in the mass-balance equation and how it affects the reactor design.

Example 4.1: Mass-Balance Equation: Air Dilution (No Chemical Reaction Occurring)

A glass bottle containing 900 mL of methylene chloride (CH_2Cl_2, specific gravity = 1.335) was accidentally left uncapped in a poorly ventilated room (5 m × 6 m × 3.6 m) over a weekend. On the following Monday it was found that two-thirds of methylene chloride had volatilized. An exhaust fan ($Q = 200$ ft^3/min) was turned on to vent the fouled air out of the laboratory. How long will it take to reduce the concentration down below the Occupational Safety and Health Administration (OSHA's) short-term exposure limit (STEL) of 125 ppmV?

Strategy:

This is a special case (no reactions occurring) of the general mass-balance equation. For this case Equation 4.5 can be simplified into:

$$V\frac{dC}{dt} = \sum Q_{in}C_{in} - \sum Q_{out}C_{out} \qquad (4.6)$$

The equation can be further simplified with the following assumptions:

1. The air leaving the laboratory is only through the exhaust fan, and the air ventilation rate is equal to the rate of air entering the laboratory ($Q_{in} = Q_{out} = Q$).
2. The air entering the laboratory does not contain methylene chloride ($C_{in} = 0$).
3. The air in the laboratory is fully mixed; thus the concentration of methylene chloride in the laboratory is uniform and is the same as that of the air vented by the fan ($C = C_{out}$).

$$V\frac{dC}{dt} = -QC \qquad (4.7)$$

It is a first-order differential equation. It can be integrated with the initial condition (i.e., $C = C_0$ at $t = 0$):

$$\frac{C}{C_0} = e^{-(Q/V)t} \quad \text{or} \quad C = C_0\, e^{-(Q/V)t} \tag{4.8}$$

Solution:

(a) Methylene chloride concentration in the laboratory before ventilation can be found as 2,100 ppmV (see Example 2.4 for detailed calculations).

(b) The size of the reactor (V) = the size of the laboratory
$$= (5 \text{ m})(6 \text{ m})(3.6 \text{ m}) = 108 \text{ m}^3$$

The system flow rate (Q) = ventilation rate
$$= 200 \text{ ft}^3/\text{min} = (200 \text{ ft}^3/\text{min})$$
$$\div (35.3 \text{ ft}^3/\text{m}^3) = 5.66 \text{ m}^3/\text{min}$$

The initial concentration, $C_0 = 2,100$ ppmV

The final concentration, $C = 125$ ppmV
$$125 = (2,100)e^{-(5.66/108)t}$$

Thus, $t = 53.8$ min

Discussion:

The actual time required would be longer than 58 min, because the assumption of completely mixed air inside the room may not be valid. In addition, if the ambient air contains some methylene chloride, the cleanup time would be even longer.

4.3 Chemical Kinetics

Chemical kinetics provides information on the rate at which a chemical reaction occurs. This section discusses the rate equation, reaction-rate constant, and reaction order. *Half-life*, a term commonly used with regard to the fate of COCs in the environment, is also described.

4.3.1 Rate Equations

In addition to the mass-balance equation, the reaction-rate equation is another relationship required for design of a homogeneous reactor. The following general mathematical expression describes the rate that the concentration of species A (C_A) changes with time:

$$\gamma_A = \frac{dC_A}{dt} = -kC_A{}^n \qquad (4.9)$$

where n is the order of the reaction, k is the reaction-rate constant, and γ_A is the rate of conversion of species A. If n is equal to 1, it is called a first-order reaction. It implies that the reaction rate is proportional to the concentration of the species. In other words, the higher the compound concentration, the faster is the reaction rate. The first-order kinetics is applicable in many environmental engineering applications. Consequently, discussion in this book will be focused on the first-order reactions and their applications. A first-order reaction can then be written as:

$$\gamma_A = \frac{dC_A}{dt} = -kC_A \qquad (4.10)$$

The rate constant itself provides lots of valuable information with regard to the reaction. A larger k value implies a faster reaction rate, which, in turn, demands a smaller reactor volume to achieve a specific conversion. The value of k varies with temperature. In general, the higher the temperature, the larger the k value will be for a reaction.

What would be the units of a reaction-rate constant for a first-order reaction? Let us take a close look at Equation (4.10). In that equation, the units of dC_A/dt is concentration/time and that of C is concentration; therefore, the units of k should be 1/time. Consequently, if a reaction-rate constant has a value of 0.25 day^{-1}, the reaction should be a first-order reaction. The units of k for zeroth-order reactions and second-order reactions are [(concentration)/time] and [(concentration)(time)]$^{-1}$, respectively.

Equation (4.10) tells us that the concentration of compound A is changing with time. This equation can be integrated between $t = 0$ and time t:

$$\ln \frac{C_A}{C_{A0}} = -kt \quad \text{or} \quad \frac{C_A}{C_{A0}} = e^{-kt} \qquad (4.11)$$

where C_{A0} is the concentration of compound A at $t = 0$, and C_A is the concentration at time t.

Example 4.2: Estimate the Rate Constant from Two Known Concentration Values (1)

An accidental gasoline spill occurred at a site 20 days ago. The total petroleum hydrocarbon (TPH) concentration at a specific location in soil dropped from an initial 3,000 mg/kg to the current 2,750 mg/kg. The decrease in

concentration is mainly attributed to natural biodegradation and volatilization. Assume that both removal mechanisms are first-order reactions and
that the reaction-rate constants for both mechanisms are independent of TPH
concentration and are constant. Estimate how long it will take for the concentration to drop below 100 mg/kg due to these natural attenuation processes.

Strategy:

Only the initial concentration and the concentration at day 5 are given.
We need to take a two-step approach to solve the problem: First
determine the rate constant, and then use the rate constant to determine the time needed to reach a final concentration of 100 mg/kg.

Two removal mechanisms (i.e., biodegradation and volatilization) are
occurring simultaneously, and both of them are first-order. These
two mechanisms are additive, and they can be represented by one
single equation with a combined rate constant.

$$\frac{dC}{dt} = -k_1 C - k_2 C = -(k_1 + k_2)C = -kC \tag{4.12}$$

Solution:

(a) Insert the initial concentration and the concentration at day 5
into Equation (4.11) to obtain k:

$$\ln\frac{2,750}{3,000} = -k(20)$$

So, $k = 0.00435$/day.

(b) For the concentration to drop below 100 mg/kg, it will take (from
Equation 4.11):

$$\ln\frac{100}{3000} = -0.00435(t)$$

$$t = 782 \text{ days}$$

**Example 4.3: Estimate the Rate Constant from Two
Known Concentration Values (2)**

The subsurface soil at a site was impacted by an accidental spill of gasoline.
A soil sample, taken 10 days after removal of the polluting source, showed
a TPH concentration of 1,200 mg/kg. The second sample taken at 25 days
showed a drop of concentration to 1,100 mg/kg. Assume that a combination
of all the removal mechanisms, including volatilization, biodegradation, and
oxidation, shows first-order kinetics. Estimate how long it will take for the

concentration to drop below 100 mg/kg without any remediation measures taken.

Strategy:

Only two concentrations at two different days are given. We need to take a two-step approach to solve the problem. We need to determine the rate constant first and then the initial concentration.

Solution:

(a) Determine the rate constant, k:

At $t = 10$ days, insert the concentration value into Equation (4.11):

$$\frac{1,200}{C_i} = e^{-k(10)}$$

At $t = 25$ days, insert the concentration value into Equation (4.11):

$$\frac{1,100}{C_i} = e^{-k(25)}$$

By dividing both sides of the first equation by the corresponding sides of the second equation, we can obtain

$$\frac{1,200}{1,100} = e^{-10k} \div e^{-25k} = e^{-10k-(-25k)} = e^{15k}$$

Thus, $k = 0.0058$/day

(b) Estimate the initial concentration (immediately after the spill)

C_i can be readily determined by inserting the value of k into either of the first two equations:

$$\frac{1,200}{C_i} = e^{-(0.0058)(10)} = 0.944$$

So, $C_i = 1,272$ mg/kg.

(c) For the concentration to drop below 100 mg/kg, it will take:

$$\frac{100}{1,272} = 0.0786 = e^{-0.0058t}$$

$t = 438$ days

4.3.2 Half-Life

The half-life can be defined as the time needed to have one-half of the COC degraded. In other words, it is the time required for the concentration to drop to half of the initial value. For first-order reactions, the half-life (often shown as $t_{1/2}$) can be found from Equation (4.11) by substituting C_A by one-half of C_{A0} (i.e., $C_A = 0.5 C_{A0}$):

$$t_{1/2} = \frac{\ln 2}{k} = \frac{0.693}{k} \tag{4.13}$$

As shown in Equation (4.13), the half-life and the rate constant are inversely proportional for the first-order reactions. If a value of half-life is given, we can find the rate constant readily from Equation (4.13), and vice versa.

Example 4.4: Half-Life Calculation (1)

The half-life of 1,1,1-trichloroethane (1,1,1-TCA) in subsurface was determined to be 180 days. Assume that all the removal mechanisms are first-order. Determine (1) the rate constant and (2) the time needed to drop the concentration down to 10% of the initial concentration.

Solution:

(a) The rate constant can be easily determined from Equation (4.13) as:

$$t_{1/2} = 180 = \frac{0.693}{k}$$

Thus, $k = 0.00385$/day.

(b) Use Equation (4.11) to determine the time needed to drop the concentration down to 10% of the initial value (i.e., $C = 0.1C_0$):

$$\frac{C}{C_0} = \frac{1}{10} = e^{-(0.00385)(t)}$$

Therefore, $t = 598$ days.

Example 4.5: Half-Life Calculation (2)

On some occasions, the decay rate is expressed as T_{90} instead of $t_{1/2}$. T_{90} is the time required for 90% of the compound to be converted (or the concentration to drop to 10% of the initial value). Derive an equation to relate T_{90} with the first-order reaction-rate constant.

Solution:

The relationship between T_{90} and k can be determined from Equation (4.11) as:

$$\frac{C}{C_i} = \frac{1}{10} = e^{-kT_{90}}$$

Then,

$$T_{90} = \frac{-\ln(0.1)}{k} = \frac{2.30}{k} \tag{4.14}$$

Example 4.6: Half-Life Calculation (3)

Methyl mercury (CH_3Hg^+) is an organic form of mercury, and it is bioaccumulative in organisms. If the metabolic process for expelling it from human body is a first-order reaction and the average excretion rate is 2% of the total body burden per day, determine the half-life of this compound in the body and how long it will take to drop the concentration in the body by 90%.

Solution:

(a)   As given, the reaction-rate constant is equal to 0.02/day; the half-life can be found from Equation (4.13):

$$t_{1/2} = \frac{0.693}{k} = \frac{0.693}{0.02} = 34.65 \text{ days}$$

(b)   The time to reach 90% reduction in the original concentration can be found from Equation (4.14):

$$T_{90} = \frac{2.30}{k} = \frac{2.30}{0.02} = 115 \text{ days}$$

Discussion:

1.   The 90% reduction is often called one log reduction.
2.   The part (b) of Example 4.4 can also be found from Equation (4.14) (i.e., $T_{90} = (2.30)/(0.02) = 115$ days).

4.4 Types of Reactors

Reactors are typically classified based on their flow characteristics and the mixing conditions within the reactor. Reactors may be operated in either a batchwise or a continuous-flow mode. In a batch reactor, the reactor is charged with the reactants, and the content is well mixed and left to react. At the end of a specified time period, the resulting mixture is discharged. A batch reactor is an unsteady-state reactor, because the composition of the reactor content changes with time. The capital cost of a batch reactor is usually less than that of a continuous-flow reactor, but it is very labor intensive, and the operating costs are higher. It is usually limited to small installations and to the cases when raw materials are expensive.

In a continuous-flow reactor, the feed to the reactor and the discharge from it are continuous. In most of the cases, the flow reactors are operated under steady-state conditions in which the feed stream flow rate, its composition, the reaction condition in the reactor, and the withdrawal rate are constant with respect to time. Frequently, reaction kinetics is studied in a laboratory to determine the rate constant by using a batch reactor. The application of the obtained rate constant to the design of a continuous-flow reactor, however, involves no changes in the kinetics principles; thus, it is valid. In general, there are two ideal types of flow reactors: continuous-flow stirred tank reactor (CFSTR) and plug-flow reactor (PFR). They are classified mainly by the mixing conditions within the reactors.

The CFSTR consists of a stirred tank that has feed stream(s) of the reactants and discharge stream(s) of reacted materials. The CFSTR is usually round, square, or slightly rectangular in a plan view, and it is necessary to provide sufficient mixing. The stirring of a CFSTR is extremely important, and it is assumed that the fluid in the reactor is perfectly mixed (i.e., the content is uniform throughout the entire reactor volume). As the result of mixing, the composition of the discharge stream(s) is the same as that of the reactor content. Therefore, it is also called a completely stirred tank reactor (CSTR) or completely mixed flow reactor (CMF). Under the steady-state conditions, the concentration of the effluent and that at any location within the reactor are the same and should not change with time.

The PFR ideally has the geometric shape of a long tube or tank and has a continuous flow in which the fluid particles pass through the reactor in series. The reactants enter at the upstream end of the reactor, and the products leave at the downstream end. Ideally, there is no induced mixing between elements of fluid along the direction of flow. Those fluid particles that enter the reactor first will leave first. The composition of the reacting fluid changes in the direction of flow. For the case of COC removal or destruction, the concentration will be the highest at the entrance and dropped continuously to the effluent value at the exit condition. Under the steady-state conditions, the effluent concentration and concentration at any location within the reactor should not change with time.

It should be noted that CFSTRs and PFRs are ideal reactors. The continuous-flow reactors in the real world behave somewhere between these ideal cases. The ideal CFSTRs are more resistant to shock loadings because the influent would be mixed with the reactor content immediately. They are a better choice if the process is sensitive to shock loadings (e.g., biological processes). On the other hand, the ideal PFRs provide the same residence time for all the influent flow. They are a better choice for chlorine contact tanks in which a minimum contact time between pathogens and disinfectants is needed. (Note: the residence times of influent parcels in an ideal CFSTR can range from extremely short to extremely long.)

4.4.1 Batch Reactors

Let us consider a batch reactor with a first-order reaction. By combining Equations (4.10) and (4.11), the mass-balance equation can be expressed as follows:

$$V\frac{dC}{dt} = (V \times \gamma) = V(-kC)$$

or   (4.15)

$$\frac{dC}{dt} = -kC$$

It is a first-order differential equation and can be integrated with the initial condition ($C = C_i$ at $t = 0$) and the final condition ($C = C_f$ at $t = $ residence time (τ)). The residence time can be defined as the time that the fluid stays inside the reactor and undergoes reaction. The integral of Equation (4.15) is

$$\frac{C_f}{C_i} = e^{-k\tau} \quad \text{or} \quad C_f = (C_i)e^{-k\tau}$$ (4.16)

Table 4.1 tabulates the design equations for batch reactors in which zeroth-, first-, and second-order reactions take place.

TABLE 4.1

Design Equations for Batch Reactors

Order of Reaction	Design Equation	Equation No.
0	$C_f = C_i - k\tau$	(4.17)
1	$C_f = C_i(e^{-k\tau})$	same as Equation (4.16)
2	$C_f = \dfrac{C_i}{1 + (k\tau)C_i}$	(4.18)

Example 4.7: Batch Reactor (Determine the Required Residence Time with a Known Rate Constant)

A batch reactor is to be designed to treat soil containing 200 mg/kg of polychlorinated biphenyls (PCBs). The required removal, conversion, or reduction of PCBs is 90%. The rate constant is 0.5 h^{-1}. What is the required residence time for this batch reactor? What is the required residence time if the desired final concentration is 10 mg/kg?

Strategy:

Although the order of the reaction is not mentioned in the problem statement, it is a first-order reaction because the units of k are 1/time.

Solution:

(a) For a 90% reduction ($\eta = 90\%$)

$$C_f = C_i (1 - \eta)$$
$$= 200 (1 - 90\%) = 20 \text{ mg/kg}$$

Insert the known values into Equation (4.16)

$$\frac{20}{200} = 0.1 = e^{-(0.5)\tau}$$

$\tau = 4.6 \text{ h}$

(b) To achieve a final concentration of 10 mg/kg:

$$\frac{10}{200} = 0.05 = e^{-(0.5)\tau}$$

$\tau = 6.0 \text{ h}$

Example 4.8: Batch Reactor (Determine the Required Residence Time with an Unknown Rate Constant)

A batch reactor was installed to remediate soil impacted by PCBs. A test run was conducted with an initial PCB concentration of 250 mg/kg. After 10 hours of batchwise operation, the concentration dropped to 50 mg/kg. However, it is required to reduce the concentration down to 10 mg/kg. Determine the required residence time to achieve the final concentration of 10 mg/kg.

Strategy:

It requires a two-step approach to solve this problem. The first is to determine the rate constant using the given information. Then, use this obtained k value to estimate the residence time for other conversions. The given information did not tell us the order of the reaction. We assume it is a first-order reaction, but this should be confirmed with additional test data.

Solution:

(a) Insert the known values into Equation (4.16) to find the value of k:

$$\frac{50}{250} = 0.20 = e^{-k(10)}$$

$$k = 0.161 \text{ h}^{-1}$$

(b) The time required to achieve a concentration of 10 mg/kg:

$$\frac{10}{250} = 0.04 = e^{-0.161\tau}$$

$$\tau = 20.0 \text{ h}$$

Discussion:

It is assumed that the reaction is first-order. One should check the validity of this assumption, for example, by running the pilot run longer. For example, if the run is extended to 20 hours and the final concentration is close to 10 mg/kg, the assumption of first-order kinetics should be valid.

Example 4.9: Determine the Rate Constant from Batch Experiments

An in-vessel bioreactor is designed to remediate soil impacted by cresol. A bench-scale batch reactor was set up to determine the order and rate constant of the reaction. The following concentrations of cresol in the batch reactor at various times were observed and recorded as:

Time (hours)	Cresol concentration (mg/kg)
0	350
0.5	260
1	200
2	100
5	17

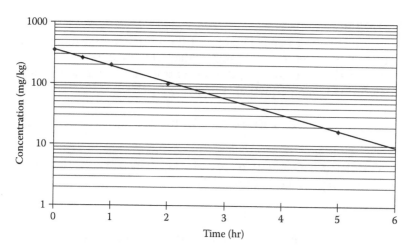

FIGURE 4.1
Concentration versus time.

Use these data to determine the reaction order and the value of the rate constant.

Strategy:

To determine the reaction order, a trial-and-error approach is often taken. From Table 4.1, if it is a zeroth-order reaction, the plot of concentration versus time should be a straight line. The plot of ln(C) versus time should be a straight line if it is first-order kinetics. If it is second-order, the plot of (1/C) versus time will be a straight line. The value of k is then obtained from the slope of the straight line.

Solution:

Since many reactions of environmental concern are first-order reactions, first assume that it is first-order and plot the concentration-time data on a semilog scale (Figure 4.1).

A straight line fits the data very well, so the assumption of first-order kinetics is valid. The slope of the straight line can be determined as 0.263/h. It should be noted that the rate constant in Equation (4.11) is based on exponential with base e, and the plot in the figure is on $\log_{10}$. Consequently, the value of k to be used in Equation (4.11) should be the product of the slope from the semilog$_{10}$ plot and 2.303 (which is the natural log of 10). That is

$$k = (0.263)(2.303) = 0.606/h$$

See Figure 4.1.

Discussion:

Using the obtained rate constant and the initial concentration to calculate the concentration at some other time t can serve as a check on the value. For example, the concentration, at $t = 2$ hours, can be determined using Equation (4.16) as:

$$C_f = 350 \times e^{-(0.606)(2)} = 104 \text{ mg/kg}$$

The calculated concentration, 104 mg/kg, is reasonably close to the reported experimental value, 100 mg/kg.

Example 4.10: Batch Reactor with Second-Order Kinetics

A batch reactor is to be designed to treat soil that contains 200 mg/kg of PCBs. The required reduction of PCBs is 90%. The rate constant is $0.5[(\text{mg}/\text{kg})(\text{h})]^{-1}$. What is the required residence time for the batch reactor?

Strategy:

Although the order of the reaction is not mentioned in the problem statement, it is a second-order reaction because the units of k are $[(\text{mg}/\text{kg})(\text{h})]^{-1}$.

Solution:

(a) For a 90% reduction ($\eta = 90\%$)

$$C_f = 200 \ (1 - 90\%) = 20 \text{ mg/kg}$$

(b) Insert the known values into Equation (4.18) (see Table 4.1):

$$20 = \frac{200}{1 + (0.5\tau)200}$$

$$\tau = 0.09 \text{ h}$$

Discussion:

The only difference between the reactors in Examples 4.7 and 4.10 is the reaction kinetics. With the same numerical value of the reaction-rate constants, the required residence time to achieve the same conversion rate is much shorter in the reactor with second-order kinetics.

TABLE 4.2

Design Equations for CFSTRs

Order of Reaction	Design Equation	Equation No.
0	$C_{out} = C_{in} - k\tau$	(4.21)
1	$\dfrac{C_{out}}{C_{in}} = \dfrac{1}{1+k\tau}$	same as Equation (4.20)
2	$\dfrac{C_{out}}{C_{in}} = \dfrac{1}{1+(k\tau)C_{out}}$	(4.22)

4.4.2 CFSTRs

Let us now consider a steady-state CFSTR with a first-order reaction. As mentioned previously, by definition, the concentration in the effluent from a CFSTR is the same as that in the reactor; and the concentration in the reactor is uniform and constant. Under steady-state conditions, the flow rate is constant, and $Q_{in} = Q_{out}$. By inserting Equation (4.10) into Equation (4.5), the mass-balance equation can be expressed as follows:

$$0 = QC_{in} - QC_{out} + (V)(-kC_{reactor})$$
$$= QC_{in} - QC_{out} + (V)(-kC_{out}) \tag{4.19}$$

With a simple mathematical manipulation, Equation (4.19) can be rearranged as:

$$\frac{C_{out}}{C_{in}} = \frac{1}{1+k(V/Q)} = \frac{1}{1+k\tau} \tag{4.20}$$

Table 4.2 tabulates the design equations for CFSTRs in which zeroth-, first-, and second-order reactions take place.

Example 4.11: A Soil Slurry Reactor with First-Order Kinetics (CFSTR)

A soil slurry reactor is used to treat soil that contains 1,200 mg/kg of TPH. The required final soil TPH concentration is 50 mg/kg. From a bench-scale study, the rate equation was found to be

$$\gamma = -0.25C \quad \text{in mg/kg/min}$$

The content in the reactor is fully mixed. Assume that the reactor behaves as a CFSTR. Determine the required residence time.

Strategy:

It is a first-order reaction, and the reaction-rate constant is equal to 0.25/min.

Solution:

Insert the known values into Equation (4.20) to find out the value of τ:

$$\frac{C_{out}}{C_{in}} = \frac{50}{1200} = \frac{1}{1+0.25\tau}$$

$\tau = 92$ min

Example 4.12: A Low-Temperature Thermal Desorption Reactor with Second-Order Kinetics (CFSTR)

A low-temperature thermal desorption reactor is used to treat soil that contains 2,500 mg/kg of TPH. The required final soil TPH concentration is 100 mg/kg. From a bench-scale study, the rate equation was found to be

$$\gamma = -0.12\,C^2 \quad \text{in mg/kg/h}$$

The reactor is rotated to achieve a good mixing. Assume that the reactor behaves as a CFSTR. Determine the required residence time.

Strategy:

It is a second-order reaction, and the reaction-rate constant is equal to 0.12/(mg/kg/h).

Solution:

Insert the known values into Equation (4.22) (see Table 4.2) to find out the value of τ:

$$\frac{C_{out}}{C_{in}} = \frac{100}{1200} = \frac{1}{1+0.12\tau(100)}$$

$\tau = 0.92$ h = 55 min

4.4.3 PFRs

Let us now consider a steady-state PFR with a first-order reaction. As mentioned previously, by definition, there is no longitudinal mixing within the

PFR. The concentration in the reactor ($C_{reactor}$) decreases from C_{in} at the inlet to C_{out} at the exit. Under the steady-state condition, the flow rate is constant, and $Q_{in} = Q_{out}$. By inserting Equation (4.10) into Equation (4.5), the mass-balance equation can be expressed as follows:

$$0 = QC_{in} - QC_{out} + (V)(-kC_{reactor}) \tag{4.23}$$

$C_{reactor}$ is a variable. The equation can be solved by considering an infinitesimal section of the reactor and integrating the equation. The solution can be expressed as follows:

$$\frac{C_{out}}{C_{in}} = e^{-k(V/Q)} = e^{-k\tau} \tag{4.24}$$

Table 4.3 tabulates the design equations for PFRs in which zeroth-, first-, and second-order reactions take place.

When comparing the design equations for PFRs in Table 4.3 and those for CFSTRs in Table 4.2, the following remarks can be derived:

Zeroth-order reactions: The design equations are identical for both reactor types. This means that the conversion rate is independent of the reactor types, provided all the other conditions are the same.

First-order reactions: The ratio of the effluent and influent concentrations is linearly proportional to the inverse of the residence time for CFSTRs, while this ratio is exponentially proportional to the inverse of the residence time for PFRs. In other words, the effluent concentration from PFRs decreases more sharply with the increase of the residence time than that from CFSTRs provided all the other conditions are the same. We can also say that for a given residence time (or reactor size), the effluent concentration from a PFR would be lower than that from a CFSTR. (More discussion and examples will be given later in this section.)

TABLE 4.3

Design Equations for PFRs

Order of Reaction	Design Equation	Equation No.
0	$C_{out} = C_{in} - k\tau$	(4.25)
1	$C_{out} = C_{in}(e^{-k\tau})$	same as Equation (4.24)
2	$C_{out} = \dfrac{C_{in}}{1 + (k\tau)C_{in}}$	(4.26)

Second-order reactions: The design equations for the second-order reactions are similar in format for PFRs and CFSTRs. The only difference is that the C_{out} in the denominator on the right-hand side of Equation (4.22) (see Table 4.2) is replaced by C_{in} in Equation (4.26) (see Table 4.3). Since $C_{in} > C_{out}$, the ratio of C_{out}/C_{in} of a PFR will be smaller than that of a CFSTR for the same C_{in}, k, and τ. The smaller C_{out}/C_{in} ratio means that the effluent concentration from a PFR would be lower than that from a CFSTR for the same C_{in}, k, and τ.

Example 4.13: A Soil Slurry Reactor with First-Order Kinetics (PFR)

A soil slurry reactor is used to treat soil that contains 1,200 mg/kg of TPH. The required final soil TPH concentration is 50 mg/kg. From a bench-scale study, the rate equation was found to be

$$\gamma = -0.25C \quad \text{in mg/kg/min}$$

Assume that the reactor behaves as a PFR. Determine the required residence time.

Strategy:
It is a first-order reaction, and the reaction-rate constant is equal to 0.25/min.

Solution:
Insert the known values into Equation (4.24) to find out the value of τ:

$$\frac{C_{out}}{C_{in}} = \frac{50}{1200} = e^{-(0.25)\tau}$$

$$\tau = 12.7 \text{ min}$$

Discussion:

1. For the same inlet concentration and reaction-rate constant, the required residence time to achieve a specified final concentration is 12.7 min for a PFR, which is much shorter than that for a CFSTR, 92 min (Example 4.11).

2. For the first-order kinetics, the reaction rate is proportional to the concentration inside the reactor (i.e., $\gamma = kC_{reactor}$). The higher the reactor concentration, the faster is the reaction rate. For CFSTRs, by definition, the reactor concentration is equal to the effluent concentration (i.e., 50 mg/kg in this case). For PFRs, by definition, the reactor concentration decreases from C_{in} (1,200 mg/kg) at the inlet to C_{out} (50 mg/kg) at the outlet. The average concentration

in the PFR (625 mg/kg as the arithmetic average or 245 mg/kg as the geometric average) is much higher than 50 mg/kg, which makes the reaction rate much higher. Consequently, the required residence time would be much shorter.

Example 4.14: A Low-Temperature Thermal Desorption Reactor with Second-Order Kinetics (PFR)

A low-temperature thermal desorption reactor is used to treat soil that contains 2,500 mg/kg of TPH. The required final soil TPH concentration is 100 mg/kg. From a bench-scale study, the rate equation was found to be

$$\gamma = -0.12 \, C^2 \quad \text{in mg/kg/hr}$$

The soil is carried through the reactor on a conveyor belt. Assume that the reactor behaves as a PFR. Determine the required residence time.

Strategy:

It is a second-order reaction, and the reaction-rate constant is equal to 0.12/(mg/kg/h).

Solution:

Insert the known values into Equation (4.26) (see Table 4.3) to find out the value of τ:

$$\frac{C_{out}}{C_{in}} = \frac{100}{1,200} = \frac{1}{1 + 0.12\tau(1,200)}$$

$$\tau = 0.08 \, h = 4.8 \, min$$

Discussion:

Again, for the same initial concentration and reaction-rate constant, the required residence time to achieve the specified final concentration is 4.8 min for a PFR, which is much shorter than that for a CFSTR, 55 min (as shown in Example 4.12).

4.5 Sizing the Reactors

Once the reactor type is selected and the required residence time to achieve the desired removal is determined, sizing a reactor is straightforward. The longer the compound needs to stay in a reactor to achieve the desired removal, the larger the reactor would be for a given flow rate.

For flow reactors such as CFSTRs and PFRs, the residence time, or the hydraulic detention time, τ, can be defined as:

$$\tau = \frac{V}{Q} \tag{4.27}$$

where V is the volume of the reactor and Q is the flow rate. For a PFR, by definition, each fluid particle should spend exactly the same amount of time flowing through the reactor. On the other hand, for a CFSTR, most fluid particles would flow through the reactor in a shorter or longer time than the average residence time. Therefore, the value of τ in Equation (4.27) is the average hydraulic detention time, which is used in determining the size of the reactor.

For a batch reactor, the residence time calculated from Equations (4.16), (4.17), and (4.18) (see Table 4.1) is the actual time needed for reaction to be accomplished. For system operation and design, an engineer needs to take the time needed for load, unload, and idle into consideration.

Example 4.15: Sizing a Batch Reactor

A soil slurry batch reactor is used to treat soils that contain 1,200 mg/kg of TPH. It is required to treat the slurry at 30 gal/min. The required final soil TPH concentration is 50 mg/kg. From a bench-scale study, the rate equation was found to be

$$\gamma = -0.05C \quad \text{in mg/kg/min}$$

The time required for loading and unloading of the slurry for each batch is 2 h. Size the batch reactor for this project.

Strategy:

It is a first-order reaction, and the reaction-rate constant is equal to 0.05/min.

Solution:

(a) Insert the known values into Equation (4.16) to find out the value of τ:

$$\frac{C_{out}}{C_{in}} = \frac{50}{1200} = e^{-(0.05)\tau}$$

$\tau = 64$ min (needed for reaction)

(b) The total time needed for each batch

= reaction time + time for loading and unloading

= 64 + 120 = 184 min

(c)   The required reactor volume, $V = (\tau)Q$ (from Equation 4.27)

$= (64 \text{ min})(30 \text{ gal/min}) = 1{,}920 \text{ gal}$

Discussion:

A minimum of three reactors are needed in this case. The reactors are operated in different phases; while two are in loading or unloading phases, the other one will be in active reaction phase. Consequently, the influent flow will not be interrupted.

Example 4.16: Sizing a CFSTR

A soil slurry reactor is used to treat soils that contain 1,200 mg/kg of TPH. It is required to treat the slurry at 30 gal/min. The required final soil TPH concentration is 50 mg/kg. From a bench-scale study, the rate equation was found to be

$$\gamma = -0.05C \quad \text{in mg/kg/min}$$

The contents in the reactor are fully mixed. Assume that the reactor behaves as a CFSTR. Size the CFSTR for this project.

Solution:

(a)   Insert the known values into Equation (4.20) to find the value of τ:

$$\frac{C_{out}}{C_{in}} = \frac{50}{1200} = \frac{1}{1+(0.05)\tau}$$

$\tau = 460 \text{ min}$

(b)   The required reactor volume, $V = (\tau)Q$ (from Equation 4.27)

$= (460 \text{ min})(30 \text{ gal/min}) = 13{,}800 \text{ gal}$

Example 4.17: Sizing a PFR

A soil slurry reactor is used to treat soil that contains 1,200 mg/kg of TPH. It is required to treat the slurry at 30 gal/min. The required final soil TPH concentration is 50 mg/kg. From a bench-scale study, the rate equation was found to be

$$\gamma = -0.05C \quad \text{in mg/kg/min}$$

Assume that the reactor behaves as a PFR. Size the PFR for this project.

Solution:

(a) Insert the known values into Equation (4.24) to find out the value
of τ:

$$\frac{C_{out}}{C_{in}} = \frac{50}{1200} = e^{-(0.05)\tau}$$

$\tau = 64$ min

(b) The required reactor volume, $V = (\tau)Q$ (from Equation 4.24)

$= (64 \text{ min})(30 \text{ gal/min}) = 1{,}920 \text{ gal}$

Discussion:

1. To achieve the same conversion, the size of the PFR, 1,920 gallons
(this example), is much smaller than 13,800 gallons for the CFSTR
(Example 4.16).

2. The design equations for batch reactors and PFRs are essentially
the same. The required reaction times for these two reactors
are the same, at 64 min. The actual tankage of the PFR is much
smaller because loading and unloading time does not need to be
included in operation of flow reactors.

4.6 Reactor Configurations

In practical engineering applications, it is more common to have a few
smaller reactors than to have one large reactor for the following reasons:

- Flexibility (to handle fluctuations of flow rate)
- Maintenance consideration
- A higher removal efficiency

Common reactor configurations include arrangement of reactors in series, in
parallel, or a combination of both.

4.6.1 Reactors in Series

For reactors in series, the flow rates to all the reactors are the same and equal
to the influent flow rate to the first reactor, Q (Figure 4.2). The first reactor,
with a volume V_1, will reduce the influent COC concentration, C_0, and yields
an effluent concentration, C_1. The effluent concentration from the first reactor
becomes the influent concentration to the second reactor. Consequently, the

FIGURE 4.2
Three reactors in series.

effluent concentration from the second reactor, C_2, becomes the influent concentration to the third reactor. More reactors can be added in series until the effluent concentration from the last reactor in series meets the requirement. For CFSTRs, a few small reactors in series will yield a lower final effluent concentration than a large reactor with the same total volume. This will be illustrated in the examples of this section.

For three CFSTRs arranged in series, the effluent concentration from the third reactor can be determined from the COC concentration in the raw waste stream as:

$$\frac{C_3}{C_0} = \left(\frac{C_3}{C_2}\right)\left(\frac{C_2}{C_1}\right)\left(\frac{C_1}{C_0}\right) = \left(\frac{1}{1+k_3\tau_3}\right)\left(\frac{1}{1+k_2\tau_2}\right)\left(\frac{1}{1+k_1\tau_1}\right) \qquad (4.28)$$

For three PFRs arranged in series, the effluent concentration from the third reactor can be determined from the COC concentration in the raw waste stream as:

$$\frac{C_3}{C_0} = \left(\frac{C_3}{C_2}\right)\left(\frac{C_2}{C_1}\right)\left(\frac{C_1}{C_0}\right) = (e^{-k_3\tau_3})(e^{-k_2\tau_2})(e^{-k_1\tau_1}) = e^{-(k_1\tau_1+k_2\tau_2+k_3\tau_3)} \qquad (4.29)$$

Example 4.18: CFSTRs in Series

Subsurface soil at a site is impacted by diesel fuel at a concentration of 1,800 mg/kg. Aboveground remediation, using slurry bioreactors, is proposed. The treatment system is required to handle a slurry flow rate of 0.04 m³/min. The required final diesel concentration in the soil is 100 mg/kg. The reaction is first-order with a rate constant 0.1/min, as determined from a bench-scale study.

Four different configurations of slurry bioreactors in the CFSTR mode are considered. Determine the final effluent concentration from each of these arrangements and whether it meets the cleanup requirement:

(a) One 4-m³ reactor
(b) Two 2-m³ reactors in series
(c) One 1-m³ reactor followed by one 3-m³ reactor
(d) One 3-m³ reactor followed by one 1-m³ reactor

Solution:

(a) For the 4-m^3 reactor,

the residence time = V/Q = 4 m^3/(0.04 m^3/min) = 100 min

Use Equation (4.20) to find the final effluent concentration:

$$\frac{C_{out}}{C_{in}} = \frac{C_{out}}{1,800} = \frac{1}{1+(0.1)(100)}$$

C_{out} = 164 mg/kg (It exceeds the cleanup level.)

(b) For the two 2-m^3 reactors,

the residence time = V/Q = 2 m^3/(0.04 m^3/min) = 50 min each

Use Equation (4.28) to find the final effluent concentration:

$$\frac{C_2}{C_0} = \left(\frac{C_2}{1,800}\right) = \left(\frac{C_2}{C_1}\right)\left(\frac{C_1}{C_0}\right) = \left[\frac{1}{1+(0.1)(50)}\right] \times \left[\frac{1}{1+(0.1)(50)}\right]$$

C_{out} = 50 mg/kg (It is below the cleanup level.)

(c) The residence time of the first reactor = 1 m^3/(0.04 m^3/min)
= 25 min

The residence time of the second reactor
= 3 m^3/(0.04 m^3/min) = 75 min

Use Equation (4.28) to find the final effluent concentration:

$$\frac{C_2}{C_0} = \left(\frac{C_2}{1,800}\right) = \left(\frac{C_2}{C_1}\right)\left(\frac{C_1}{C_0}\right) = \left[\frac{1}{1+(0.1)(25)}\right] \times \left[\frac{1}{1+(0.1)(75)}\right]$$

C_{out} = 60.5 mg/kg (It is below the cleanup level.)

(d)   The residence time of the first reactor = 3 m³/(0.04 m³/min)
$$= 75 \text{ min}$$

The residence time of the second reactor
$$= 1 \text{ m}^3/(0.04 \text{ m}^3/\text{min}) = 25 \text{ min}$$

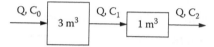

Use Equation (4.28) to find the final effluent concentration:

$$\frac{C_2}{C_0} = \left(\frac{C_2}{1,800}\right) = \left(\frac{C_2}{C_1}\right)\left(\frac{C_1}{C_0}\right) = \left[\frac{1}{1+(0.1)(75)}\right] \times \left[\frac{1}{1+(0.1)(25)}\right]$$

$C_{out} = 60.5$ mg/kg (It is below the cleanup level.)

Discussion:

1. The total volume of the reactor(s) for each of the four configurations is 4 m³.

2. The effluent concentration from the first setup (one large reactor) is the highest. Actually, having a series of smaller CFSTRs in series will always be more efficient in conversion than having a single large CFSTR. A PFR can be viewed as an infinite series of small CFSTRs, and a PFR is always more efficient than a CFSTR of equal size.

3. For the configurations having two small reactors in series, the setup with two equal-size reactors yields the lowest effluent concentration.

4. For two reactors of different sizes, the sequence of the reactors does not affect the final effluent concentration, provided that the rate constants in the reactors are the same.

Example 4.19: PFRs in Series

Subsurface soil at a site is impacted by diesel fuel at a concentration of 1,800 mg/kg. Aboveground remediation, using slurry bioreactors, is proposed. The treatment system is required to handle a slurry flow rate of 0.04 m³/min. The required final diesel concentration in the soil is 100 mg/kg. The reaction is first-order with a rate constant 0.1/min, as determined from a bench-scale study.

Four different configurations of slurry bioreactors in the PFR mode are considered. Determine the final effluent concentration from each of these arrangements and whether it meets the cleanup requirement:

(a) One 4-m^3 reactor
(b) Two 2-m^3 reactors in series
(c) One 1-m^3 reactor followed by one 3-m^3 reactor
(d) One 3-m^3 reactor followed by one 1-m^3 reactor

Solution:

(a) For the 4-m^3 reactor,

the residence time = V/Q = 4 m^3/(0.04 m^3/min) = 100 min

Use Equation (4.24) to find the final effluent concentration:

$$\frac{C_{out}}{C_{in}} = \frac{C_{out}}{1,800} = e^{-(0.1)(100)} = e^{-10}$$

C_{out} = 0.082 mg/kg (It is below the cleanup level.)

(b) For the two 2-m^3 reactors,

the residence time = V/Q = 2 m^3/(0.04 m^3/min) = 50 min each

Use Equation (4.29) to find the final effluent concentration:

$$\frac{C_2}{C_0} = \left(\frac{C_2}{1,800}\right) = \left(\frac{C_2}{C_1}\right)\left(\frac{C_1}{C_0}\right) = [e^{-(0.1)(50)}] \times [e^{-(0.1)(50)}]$$

$$= e^{-(0.1)(50+50)} = e^{-10}$$

C_{out} = 0.082 mg/kg (It is below the cleanup level.)

(c) The residence time of the first reactor

= 1 m^3/(0.04 m^3/min) = 25 min

The residence time of the second reactor
$= 3 \text{ m}^3/(0.04 \text{ m}^3/\text{min}) = 75 \text{ min}$

Use Equation (4.29) to find the final effluent concentration:

$$\frac{C_2}{C_0} = \left(\frac{C_2}{1,800}\right) = \left(\frac{C_2}{C_1}\right)\left(\frac{C_1}{C_0}\right) = [e^{-(0.1)(25)}] \times [e^{-(0.1)(75)}]$$

$$= e^{-(0.1)(25+75)} = e^{-10}$$

$C_{out} = 0.082 \text{ mg/kg}$ (It is below the cleanup level.)

(d)   The residence time of the first reactor $= 3 \text{ m}^3/(0.04 \text{ m}^3/\text{min})$
$= 75 \text{ min}$

The residence time of the second reactor
$= 1 \text{ m}^3/(0.04 \text{ m}^3/\text{min}) = 25 \text{ min}$

Use Equation (4.29) to find the final effluent concentration:

$$\frac{C_2}{C_0} = \left(\frac{C_2}{1,800}\right) = \left(\frac{C_2}{C_1}\right)\left(\frac{C_1}{C_0}\right) = [e^{-(0.1)(75)}] \times [e^{-(0.1)(25)}]$$

$$= e^{-(0.1)(75+25)} = e^{-10}$$

$C_{out} = 0.082 \text{ mg/kg}$ (It is below the cleanup level.)

Discussion:

1.   The total volume of the reactor(s) for each of the four configurations is 4 m³.
2.   The effluent concentrations from all four different configurations are the same.
3.   The effluent concentration of PFRs is much lower than those of CFSTRs in Example 4.18.

Example 4.20: CFSTRs in Series

Low-temperature thermal desorption reactors (assuming they are ideal CFSTRs) are used to treat soil that contains 1,050 mg/kg of TPH. The required final soil TPH concentration is 10 mg/kg. A reactor with a 20-min residence time can only reduce the concentration to 50 mg/kg. Assume that this is a first-order reaction. Can two smaller reactors (10-min residence time each) in series reduce the TPH concentration below 10 mg/kg?

Strategy:

The reaction-rate constant was not given, so we have to find its value first.

Solution:

(a) Use Equation (4.20) to find the rate constant:

$$\frac{C_{out}}{C_{in}} = \frac{50}{1,050} = \frac{1}{1+(k)(20)}$$

$k = 1/\text{min}$

(b) For two small reactors in series:

Use Equation (4.28) to find out the final effluent concentration,

$$\frac{C_2}{C_0} = \left(\frac{C_2}{1,050}\right) = \left(\frac{C_2}{C_1}\right)\left(\frac{C_1}{C_0}\right) = \left(\frac{1}{1+(1)(10)}\right)\left(\frac{1}{1+(1)(10)}\right)$$

$C_{out} = 8.7$ mg/kg (It is below the cleanup level.)

Discussion:

This example again demonstrates that two smaller CFSTRs can do a better job than a larger CFSTR with an equivalent total volume. However, two reactors may require a larger capital investment (two sets of process control, for example) and higher operating and maintenance (O&M) costs.

Example 4.21: PFRs in Series

UV/ozone treatment is selected to remove trichloroethylene (TCE) from an extracted groundwater stream (TCE concentration = 200 ppb). At a design flow rate of 50 L/min, an off-the-shelf reactor would provide a hydraulic retention of 5 min and reduce TCE concentration from 200 to 16 ppb. However, the discharge limit for TCE is 3.2 ppb. Assuming the reactors are of ideal plug flow type and the reaction is first-order, how many reactors would you recommend? What would be the TCE concentration in the final effluent?

Solution:

(a) Use Equation (4.24) to find out the reaction-rate constant:

$$\frac{C_{out}}{C_{in}} = \frac{16}{200} = e^{-(k)(5)}$$

$k = 0.505/min$

(b) Use Equation (4.29) to find out the final effluent concentration from two PFRs in series:

$$\frac{C_2}{C_0} = \left(\frac{C_2}{200}\right) = \left(\frac{C_2}{C_1}\right)\left(\frac{C_1}{C_o}\right) = (e^{-(0.505)(5)})(e^{-(0.505)(5)}) = e^{-5.05}$$

$C_{out} = 1.28$ ppb (It is less than 3.2 ppb.)

Two PFRs, each with 5-min residence time, would be needed.

Discussion:

We can also determine the total residence time needed to reduce the final concentration to 3.2 ppb first, and then determine the number of PFRs needed.

Use Equation (4.24) to find out the required residence time

$$\frac{C_{out}}{C_{in}} = \frac{3.2}{200} = e^{-(0.505)\tau}$$

$\tau = 8.2$ min (Two PFRs are needed.)

4.6.2 Reactors in Parallel

For reactors in parallel, the reactors share the same influent (the influent is split and fed to the reactors). The flow rate to each reactor in parallel can be

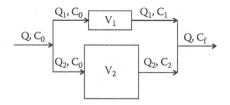

FIGURE 4.3
Two reactors in parallel.

different; however, the influent concentrations to all the reactors in parallel are the same. The sizes of the reactors may not be the same, and the effluent concentrations from the reactors can be different (Figure 4.3). In that figure, the following mass-balance equations are valid:

$$Q = Q_1 + Q_2 \tag{4.30}$$

$$C_f = \frac{Q_1 C_1 + Q_2 C_2}{Q_1 + Q_2} \tag{4.31}$$

Reactors in parallel configurations are often used in the following cases: (1) a single reactor cannot handle the flow rate; (2) the total influent rate fluctuates significantly; or (3) the reactors require frequent maintenance.

Example 4.22: CFSTRs in Parallel

Subsurface soil at a site is impacted by diesel fuel at a concentration of 1,800 mg/kg. Aboveground remediation, using slurry bioreactors, is proposed. The treatment system is required to handle a slurry flow rate of 0.04 m³/min. The required final diesel concentration in the soil is 100 mg/kg. The reaction is first-order with a rate constant 0.1/min, as determined from a bench-scale study.

Four different configurations of slurry bioreactors in the CFSTR mode are considered. Determine the final effluent concentration from each of these arrangements and whether it meets the cleanup requirement:

(a) One 4-m³ reactor
(b) Two 2-m³ reactors in parallel (each receives 0.02 m³/min flow)
(c) One 1-m³ reactor and one 3-m³ reactor in parallel (each receives 0.02 m³/min flow)
(d) One 1-m³ reactor and one 3-m³ reactor in parallel (the smaller reactor receives 0.01 m³/min flow while the larger one receives 0.03 m³/min flow)

Solution:

(a) For the 4-m³ reactor,

the residence time = $V/Q = 4$ m³/(0.04 m³/min) = 100 min

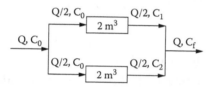

Use Equation (4.20) to find the final effluent concentration:

$$\frac{C_{out}}{C_{in}} = \frac{C_{out}}{1,800} = \frac{1}{1+(0.1)(100)}$$

$C_{out} = 164$ mg/kg (It exceeds the cleanup level.)

(b) For the two 2-m³ reactors,

the residence time = $V/Q = 2$ m³/(0.02 m³/min) = 100 min each

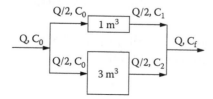

Use Equation (4.20) to find the effluent concentration from each reactor:

$$\frac{C_{out}}{C_{in}} = \frac{C_{out}}{1,800} = \frac{1}{1+(0.1)(100)}$$

$C_{out} = 164$ mg/kg for both reactors (The combined final effluent exceeds the cleanup level.)

(c) The residence time of the first reactor = 1 m³/(0.02 m³/min)

= 50 min

The residence time of the second reactor

= 3 m³/(0.02 m³/min) = 150 min

Use Equation (4.20) to find the effluent concentration from each
reactor:

$$\text{Reactor 1:} \quad \frac{C_1}{1,800} = \frac{1}{1+(0.1)(50)}$$

$C_1 = 300 \text{ mg/kg}$

$$\text{Reactor 2:} \quad \frac{C_1}{1,800} = \frac{1}{1+(0.1)(150)}$$

$C_2 = 112.5 \text{ mg/kg}$

Use Equation (4.31) to find the concentration of the combined
effluent:

$$C_f = \frac{2(300)+2(112.5)}{2+2} = 206 \text{ mg/kg}$$

$C_{out} = 206$ mg/kg for both reactors (The combined final effluent
exceeds the cleanup level.)

(d)   The residence time of the first reactor $= 1 \text{ m}^3/(0.01 \text{ m}^3/\text{min})$
$= 100 \text{ min}$

The residence time of the second reactor
$= 1 \text{ m}^3/(0.04 \text{ m}^3/\text{min}) = 100 \text{ min}$

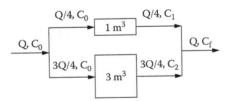

Use Equation (4.20) to find the effluent concentration from each
reactor:

$$\frac{C_{out}}{C_{in}} = \frac{C_{out}}{1,800} = \frac{1}{1+(0.1)(100)}$$

$C_{out} = 164$ mg/kg for both reactors (The combined final effluent
exceeds the cleanup level.)

Discussion:

1. The total volume of the reactor(s) for each of the four configurations is 4 m^3.

2. The effluent concentrations from all the configurations exceed the cleanup level. The configurations (a), (b), and (d) have the same effluent concentrations because the residence times of all reactors are identical. The effluent concentration from configuration (c) is the worst among the four.

3. To split the flow into reactors in parallel with the same residence time does not have any impact on the final effluent concentration, as shown in the cases (a), (b), and (d).

Example 4.23: PFRs in Parallel

Subsurface soil at a site is impacted by diesel fuel at a concentration of 1,800 mg/kg. Aboveground remediation, using slurry bioreactors, is proposed. The treatment system is required to handle a slurry flow rate of 0.04 m^3/min. The required final diesel concentration in the soil is 100 mg/kg. The reaction is first-order with a rate constant 0.1/min, as determined from a bench-scale study.

Four different configurations of slurry bioreactors in the PFR mode are considered. Determine the final effluent concentration from each of these arrangements and whether it meets the cleanup requirement:

(a) One 4-m^3 reactor

(b) Two 2-m^3 reactors in parallel (each receives 0.02 m^3/min flow)

(c) One 1-m^3 reactor and one 3-m^3 reactor in parallel (each receives 0.02 m^3/min flow)

(d) One 1-m^3 reactor and one 3-m^3 reactor in parallel (the smaller reactor receives 0.01 m^3/min flow while the larger one receives 0.03 m^3/min flow)

Solution:

(a) For the 4-m^3 reactor,

the residence time = $V/Q = 4$ m^3/(0.04 m^3/min) = 100 min

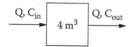

Use Equation (4.24) to find the final effluent concentration:

$$\frac{C_{out}}{C_{in}} = \frac{C_{out}}{1,800} = e^{-(0.1)(100)} = e^{-10}$$

$C_{out} = 0.082$ mg/kg (It is below the cleanup level.)

(b)   For the two 2-m^3 reactors,
the residence time = $V/Q = 2$ $m^3/(0.02$ $m^3/min) = 100$ min each

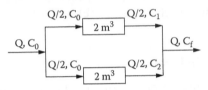

Use Equation (4.24) to find the effluent concentration for each reactor:

$$\frac{C_{out}}{C_{in}} = \frac{C_{out}}{1,800} = e^{-(0.1)(100)} = e^{-10}$$

$C_{out} = 0.082$ mg/kg for both reactors (The combined final effluent is below the cleanup level.)

(c)   The residence time of the first reactor = 1 $m^3/(0.02$ $m^3/min)$
$= 50$ min

The residence time of the second reactor
$= 3$ $m^3/(0.02$ $m^3/min) = 150$ min

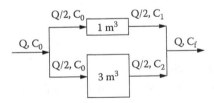

Use Equation (4.24) to find the effluent concentration from each reactor:

Reactor 1: $\dfrac{C_1}{1,800} = e^{-(0.1)(50)} = e^{-5}$

$$C_1 = 12.2 \text{ mg/kg}$$

Reactor 2: $\quad \dfrac{C_1}{1,800} = e^{-(0.1)(150)} = e^{-15}$

$C_2 = 0.00055 \text{ mg/kg}$

Use Equation (4.31) to find out the concentration of the combined effluent

$$C_f = \frac{2(12.2) + 2(5.5 \times 10^{-4})}{2+2} = 6.1 \text{ mg/kg}$$

The combined final effluent is below the cleanup level.

(d)   The residence time of the first reactor $= 1 \text{ m}^3/(0.01 \text{ m}^3/\text{min})$
$$= 100 \text{ min}$$

The residence time of the second reactor
$$= 1 \text{ m}^3/(0.04 \text{ m}^3/\text{min}) = 100 \text{ min}$$

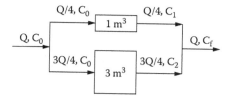

Use Equation (4.24) to find the effluent concentration from each reactor:

$$\frac{C_{out}}{C_{in}} = \frac{C_{out}}{1,800} = e^{-(0.1)(100)} = e^{-10}$$

$C_{out} = 0.082 \text{ mg/kg}$ for both reactors (The combined final effluent is below the cleanup level.)

Discussion:

1. The total volume of the reactor(s) for each of the four configurations is 4 m³.

2. The effluent concentrations from all the configurations are below the cleanup level. The configurations (a), (b), and (d) have the same effluent concentrations because the residence times of all reactors are identical. The effluent concentration from configuration (c) is the worst among the four.

3. To split the flow into reactors in parallel with the same residence time does not have any impact on the final effluent concentration, as shown in cases (a), (b), and (d).

4. Note that all the effluent concentrations from the CFSTRs in Example 4.22 exceed the cleanup level. These two examples again illustrate that PFRs are more efficient than CFSTRs, provided that the influent concentrations, reaction-rate constant, and the residence time are the same.

5

Vadose Zone Soil Remediation

5.1 Introduction

After site assessment and remedial investigation, if compounds of concern (COCs) in the subsurface at a site exceed the acceptable levels, remediation and/or removal of the impacted soil would be required. Many remedial technologies have been developed and utilized to remediate impacted soil. These technologies can be categorized into physical, chemical, biological, and thermal methods. The technologies can also be applied *in situ* and/or *ex situ*. The remedial objective is to reduce the COC concentrations below acceptable cleanup levels.

This chapter covers design calculations for some commonly used *in situ* and *ex situ* soil remediation techniques. The technologies covered include soil vapor extraction, soil washing, bioremediation, *in situ* chemical oxidation, low-temperature thermal desorption, and thermal destruction.

5.2 Soil Vapor Extraction

5.2.1 Description of the Soil-Venting Process

Soil vapor extraction (SVE), also known as soil venting, *in situ* vacuum extraction, *in situ* volatilization, or soil vapor stripping, has become a very popular remediation technique for soil impacted by volatile organic compounds (VOCs). The process strips volatile organic constituents from the impacted soil by inducing an air flow through the impacted zone. The air flow is created by a vacuum pump (often called a "blower") through a single well or a network of wells.

As the soil vapor is swept away from the void of the vadose zone, fresh air is naturally (through passive venting wells or air infiltration) or mechanically (through air-injection wells) introduced and refills the void. This flux of the fresh air will (1) disrupt the existing partition of the COCs among the void, soil moisture, and soil grain surface by promoting volatilization of the dissolved and desorption of the adsorbed COCs; (2) provide oxygen to indigenous microorganisms for biodegradation of the COCs; and (3) carry away the toxic metabolic by-products generated from the biodegradation process. The extracted air is usually laden with VOCs and brought to the ground surface by the vacuum blower. Treatment of the extracted vapor is normally required before being released to the ambient air. Design calculations for the VOC-laden air treatment are covered in Chapter 7.

Major components of a typical soil-venting system include vapor extraction well(s), vacuum blower(s), moisture-removal device (the knockout drum), off-gas collection piping and ancillary equipment, and the off-gas treatment system. The most important parameters for preliminary design of a soil-venting system are the extracted VOC concentration, air flow rate, radius of influence of the venting well, number of wells required, locations of the wells, and the size of the vacuum pump.

5.2.2 Expected Vapor Concentration

As mentioned in Section 2.4, volatile organic COCs in a vadose zone may be present in four phases: (1) in the soil moisture due to dissolution, (2) on the soil grain surface due to adsorption, (3) in the pore void due to volatilization, and (4) as the free product. If the free-product phase is present, the vapor concentration in the pore void can be estimated from Raoult's law as:

$$P_A = (P^{vap})(x_A) \tag{5.1}$$

where
P_A = partial pressure of compound A in the vapor phase
P^{vap} = vapor pressure of compound A as a pure liquid
x_A = mole fraction of compound A in the liquid phase

Examples using Raoult's law can be found in Section 2.4. The partial pressure calculated from Equation (5.1) represents the upper limit of the COC concentration in the extracted vapor from a soil-venting project. The actual concentration will be lower than this upper limit because (1) not all the extracted air passes through the impacted zone, and (2) limitations on mass transfer exist. Nevertheless, this concentration serves as a starting point for estimating the initial vapor concentration at the beginning of a soil-venting project. Initially, the extracted vapor concentrations will be relatively constant if free product is present. As soil venting continues, the free-product phase will disappear. The extracted vapor concentration will then begin to drop, and the extracted vapor concentration will become dependent on the partitioning of the COCs among the three phases. As the air flows through the pores and sweeps away the COCs, the COCs dissolved in the soil

TABLE 5.1

Physical Properties of Gasoline and Weathered Gasoline

Compound	MW (g/mole)	P^{vap} at 20°C (atm)	Saturated Vapor Concentration	
			ppmV	mg/m³
Gasoline	95	0.34	340,000	1,343,000
Weathered gasoline	111	0.049	49,000	220,000

Source: Modified from [3].

moisture have a stronger tendency to volatilize from the liquid into the void. Simultaneously, some COCs will desorb from the soil grain surface and enter into the soil moisture (assuming the soil grains are covered by a moisture layer). Consequently, the concentrations in all three phases decrease as the venting process progresses.

These phenomena describe common observations at sites that contain a single type of COC. Soil venting has also been widely used for sites impacted by a mixture of compounds, such as gasoline. For these cases, the vapor concentration decreases continuously from the start of venting; a period of constant vapor concentration in the beginning phase of the project may not exist. This can be explained by the fact that each compound in the mixture has a different vapor pressure. Thus, the more volatile compounds tend to leave the free product, as well as the moisture and the soil surface, earlier than the less volatile ones. Table 5.1 shows the molecular weights of fresh and weathered gasoline and their vapor pressures at 20°C. The table also includes the saturated vapor concentrations that are in equilibrium with the fresh and the weathered gasoline.

To estimate the initial concentration of the extracted vapor in equilibrium with the free-product phase, the following procedure can be used:

Step 1: Obtain the vapor pressure data of the COC (e.g., from Table 2.5).

Step 2: Determine the mole fraction of the COC in the free product. For a pure compound, set $x_A = 1$. For a mixture, follow the procedure in Section 2.2.4.

Step 3: Use Equation (5.1) to estimate the vapor concentration.

Step 4: Convert the concentration by volume into a mass concentration, if needed, using Equation (2.1).

Information needed for this calculation:

- Vapor pressures of the COCs
- Molecular weights of the COCs

Example 5.1: Estimate the Saturated Gasoline Vapor Concentration

Use the information in Table 5.1 to estimate the maximum gasoline vapor concentration for two soil-venting projects. Both sites are impacted by accidental gasoline spills. The spill at the first site happened recently, while the spill at the other site occurred three years ago.

Solution:

The site with fresh gasoline:

(a) Vapor pressure of fresh gasoline is 0.34 atm at 20°C, as shown in Table 5.1. The partial pressure of this gasoline in the void can be found by using Equation (5.1) as:

$$P_A = (P^{vap})(x_A) = (0.34 \text{ atm})(1.0) = 0.34 \text{ atm}$$

Thus, the partial pressure of gasoline in the air is 0.34 atm ($= 340,000 \times 10^{-6}$ atm), which is equivalent to 340,000 ppmV.

Use Equation (2.1) to convert the ppmV concentration into a mass concentration (at 20°C), as

$$1 \text{ ppmV fresh gasoline} = [(MW \text{ of fresh gasoline})/24.05] \text{ mg/m}^3$$
$$= (95)/24.05 = 3.95 \text{ mg/m}^3$$
$$\text{So, } 340,000 \text{ ppmV} = (340,000)(3.95)$$
$$= 1,343,000 \text{ mg/m}^3 = 1,343 \text{ mg/L}$$

The site with weathered gasoline:

(b) Vapor pressure of the weathered gasoline is 0.049 atm, which is equivalent to 49,000 ppmV.

Use Equation (2.1) to convert the ppmV concentration into a mass concentration (at 20°C), as

$$1 \text{ ppmV weathered gasoline} = [(MW \text{ of weathered gasoline})/24.05] \text{ mg/m}^3$$
$$= (111)/24.05 = 4.62 \text{ mg/m}^3$$
$$\text{So, } 49,000 \text{ ppmV} = (49,000)(4.62)$$
$$= 226,000 \text{ mg/m}^3 = 226 \text{ mg/L}$$

Discussion:

1. The saturated vapor concentration of the weathered gasoline can be a few times smaller than that of the fresh gasoline. In this example, it is more than five times smaller.
2. The calculated vapor concentrations are essentially the same as those listed in Table 5.1.

Example 5.2: Estimate Saturated Vapor Concentrations of a Binary Mixture

A site is impacted by an industrial solvent. The solvent consists of 50% toluene and 50% xylenes by weight. Soil venting is being considered for site remediation. Estimate the maximum vapor concentration of the extracted vapor. The subsurface temperature of the site is 20°C.

Solution:

(a) From Table 2.5, the following physicochemical properties were obtained:

$$\text{Molecular weight of toluene} = 92.1$$
$$\text{Molecular weight of xylenes} = 106.2$$
$$P^{vap} \text{ of toluene} = 22 \text{ mm-Hg}$$
$$P^{vap} \text{ of xylenes} = 10 \text{ mm-Hg}$$

(b) The mole fractions of toluene and xylenes in the solvent can be found as:

Basis: 1,000 g solvent

$$\text{Moles of toluene} = \text{mass/MW} = [(50\%)(1,000)] \div (92.1) = 5.43 \text{ moles}$$
$$\text{Moles of xylenes} = \text{mass/MW} = [(50\%)(1,000)] \div (106.2) = 4.71 \text{ moles}$$
$$\text{Mole fraction of toluene} = (5.43)/(5.43 + 4.71) = 0.536$$
$$\text{Mole fraction of xylenes} = 1 - 0.536 = 0.464$$

(c) The saturated vapor concentration can be found by using Equation (5.1) as:

$$P_{toluene} = (P^{vap})(x_A) = (22 \text{ mm-Hg})(0.536)$$
$$= 11.79 \text{ mm-Hg} = 0.0155 \text{ atm}$$

Thus, partial pressure of toluene = 0.0155 atm = 15,500 ppmV.

$$P_{xylenes} = (P^{vap})(x_A) = (10 \text{ mm-Hg})(0.464)$$

$$= 4.64 \text{ mm-Hg} = 0.0061 \text{ atm}$$

Thus, partial pressure of xylenes = 0.0061 atm = 6,100 ppmV.
The volumetric (or molar) composition of the extracted vapor
$$= (15,500)/[15,500 + 6,100] = 71.8\% \text{ (toluene)}$$

(d)   The mass concentration can be found by using Equation (2.1) as:

1 ppmV toluene = (92.1)/24.05 = 3.83 mg/m^3
So, 15,500 ppmV = (15,500)(3.83) = 59,400 mg/m^3 = 59.4 mg/L
1 ppmV xylenes = (106.2)/24.05 = 4.42 mg/m^3
So, 15,500 ppmV = (6,100)(4.42) = 27,000 mg/m^3 = 27.0 mg/L

The weight composition of the extracted vapor
$$= (59.4)/[59.4 + 27.0] = 68.8\% \text{ (toluene)}$$

Discussion:
1. The toluene concentration in the extracted vapor is 68.8% by weight and 71.8% by volume. Both are higher than its concentration in the liquid solvent, 50% by weight. The higher percentage of toluene in the vapor is mainly due to its higher vapor pressure.
2. This saturated vapor concentration would be higher than the actual concentration of the extracted vapor due to the fact that (1) not all the air flows through the impacted zone and (2) limitations on mass transfer exist.

As mentioned, the presence or absence of a free-product phase greatly affects the extracted vapor concentration. Equation (2.40) in Chapter 2 can be used as a starting point for discussion.

$$X = \left\{ \frac{[(\phi_w) + (\rho_b)K_p + (\phi_a)H]}{\rho_t} \right\} \times C$$

$$= \left\{ \frac{\left[\frac{(\phi_w)}{H} + \frac{(\rho_b)K_p}{H} + (\phi_a) \right]}{\rho_t} \right\} \times G \tag{2.40}$$

Let soil saturation concentration (X_{sat}) correspond to the COC concentration in soil at which the adsorptive limits of soil grains, the solubility of soil moisture, and the saturation of soil pore gas have been achieved. Above this saturation concentration, free product must be present. Equation (2.40) can be modified by replacing C with COC solubility in water (S_w) and G with vapor concentration in equilibrium with free product (G_{sat}) as

$$X_{sat} = \left\{ \frac{[(\phi_w)+(\rho_b)K_p+(\phi_a)H]}{\rho_t} \right\} \times S_w$$

$$= \left\{ \frac{\left[\frac{(\phi_w)}{H} + \frac{(\rho_b)K_p}{H} + (\phi_a) \right]}{\rho_t} \right\} \times G_{sat} \tag{5.2}$$

To determine if the free-product phase is present, the following procedure can be used:

Step 1: Obtain the physicochemical data of the COC (e.g., from Table 2.5).

Step 2: Assume the free-product phase is present. Use Equation (5.1) to determine the saturated vapor concentration.

Step 3: Convert the saturated vapor concentration into a mass concentration by using Equation (2.1).

Step 4: Determine the K_{oc} value using Equation (2.28) and K_p value using Equation (2.26).

Step 5: Determine the COC concentration in soil by using Equation (5.2) and the vapor concentration from Step 3 (or by using Equation (5.2) and the solubility in water).

Step 6: If the soil saturation concentration in soil determined from Step 5 is smaller than the concentrations of the soil samples, the free-product phase should be present.

Information needed for this calculation:

• Vapor pressure of the COC (or the solubility in water)
• Molecular weight of the COC
• Henry's constant of the COC
• Organic–water partition coefficient, K_{ow}
• Organic content, f_{oc}
• Porosity, ϕ
• Degree of water saturation
• Dry bulk density of soil, ρ_b
• Total bulk density of soil, ρ_t

Example 5.3: Determine if the Free-Product Phase Is Present in Subsurface Using the Saturated Vapor Concentration

A subsurface is impacted by a spill of 1,1-dichloroethane (1,1-DCA). The 1,1-DCA concentrations of the soil samples from the impacted zone were between 6,000 to 9,000 mg/kg. The subsurface has the following characteristics:

- Porosity = 0.4
- Organic content in soil = 0.02
- Degree of water saturation = 30%
- Subsurface temperature = 20°C
- Dry bulk density of soil = 1.6 g/cm³
- Total bulk density of soil = 1.8 g/cm³

Determine if the free-product phase of 1,1-DCA is present in the subsurface. What could be the maximum 1,1-DCA concentration in soil if the free-product phase of 1,1-DCA is absent?

Solution:

(a) From Table 2.5, the following physicochemical properties of 1,1-DCA were obtained:

$$\text{Molecular weight} = 99.0$$
$$H = 4.26 \text{ atm/M}$$
$$P^{vap} = 180 \text{ mm-Hg}$$
$$\log K_{ow} = 1.80$$

(b) Use Equation (5.1) to determine the saturated 1,1-DCA vapor concentration:

$$P^{vap} = 180 \text{ mm-Hg} = 0.237 \text{ atm} = 237{,}000 \text{ ppmV}$$

(c) Convert the saturated vapor concentration into a mass concentration by using Equation (2.1):

$$1 \text{ ppmV of 1,1-DCA} = (99.0)/24.05 = 4.12 \text{ mg/m}^3$$
$$\text{So, } G_{sat} = 237{,}000 \text{ ppmV} = (237{,}000)(4.12)$$
$$= 976{,}000 \text{ mg/m}^3 = 976 \text{ mg/L}$$

(d) Use Table 2.4 to convert the Henry's constant to a dimensionless value:

$$H^* = H/RT = (4.26)/[(0.082)(273+20)] = 0.177 \text{ (dimensionless)}$$

Use Equation (2.28) to find K_{oc},

$$K_{oc} = 0.63K_{ow} = 0.63 \ (10^{1.80}) = (0.63)(63.1) = 39.8$$

Use Equation (2.26) to find K_p,

$$K_p = f_{oc}K_{oc} = (0.02) \ (39.8) = 0.795 \ \text{L/kg}$$

(e)   Use Equation (5.2) to estimate the soil concentration of 1,1-DCA:

$$X_{sat} = \left\{ \frac{\left[\frac{(\varphi_w)}{H} + \frac{(\rho_b)K_p}{H} + (\phi_a) \right]}{\rho_t} \right\} \times G_{sat}$$

$$X_{sat} = \{[(0.4)(30\%)/(0.177) + (1.6)(0.795)/(0.177) + (0.4)(1 - 30\%)] \div 1.8\}$$
$$\times (976)$$
$$= 4{,}416 \ \text{mg/kg}$$

This value, 4,416 mg/kg, represents the maximum 1,1-DCA concentration in the soil if the free-product phase of 1,1-DCA is absent.

(f)   Since the calculated 1,1-DCA concentration, 4,416 mg/kg, is less than the reported concentrations of the soil samples, the free-product phase of 1,1-DCA should be present in the subsurface.

Example 5.4: Determine if the Free-Product Phase Is Present in Subsurface Using the Solubility in Water

For the site discussed in Example 5.3, determine if the free-product phase of 1,1-DCA is present in the subsurface using the solubility of 1,1-DCA in water. What could be the maximum 1,1-DCA concentration in soil if the free-product phase of 1,1-DCA is absent?

Solution:

(a)   From Table 2.5, the solubility of 1,1-DCA in water = 5,500 mg/L

(b)   Use Equation (2.37) to estimate the soil concentration of 1,1-DCA:

$$X_{sat} = \left\{ \frac{[(\phi_w) + (\rho_b)K_p + (\phi_a)H]}{\rho_t} \right\} \times S_w$$

$$X = \{[(0.4)(30\%) + (1.6)(0.795) + (0.4)(1 - 30\%)(0.177)] \div 1.8\} \times$$
$$(5{,}500)$$
$$= 4{,}405 \ \text{mg/kg}$$

This value, 4,405 mg/kg, represents the maximum 1,1-DCA concentration in the soil if the free-product phase of 1,1-DCA is
absent.

(c) Since the calculated DCA concentration, 4,405 mg/kg, is less
than the reported concentrations of the soil samples, the free-
product phase of 1,1-DCA should be present in the subsurface.

Discussion:

Estimated values of the saturated soil concentration from Example 5.3
and Example 5.4 are essentially the same.

To determine the extracted soil vapor concentration in the absence of free
product in the subsurface, the following procedures can be used:

Step 1: Obtain the physicochemical data of the COC (e.g., from Table 2.5).

Step 2: Determine the K_{oc} value using Equation (2.28) and K_p value
using Equation (2.26).

Step 3: Determine the vapor concentration by using Equation (2.40) and
the COC concentration in soil.

Information needed for this calculation:

- COC concentration of soil samples
- Henry's constant of the COC
- Organic–water partition coefficient, K_{ow}
- Organic content, f_{oc}
- Porosity, ϕ
- Degree of water saturation
- Dry bulk density of soil, ρ_b
- Total bulk density of soil, ρ_t

Example 5.5: Estimate the Extracted Vapor Concentration (in the Absence of the Free Product)

A subsurface is impacted by a benzene spill. The average benzene concentration of the soil samples, taken from the impacted zone, is 500 mg/kg. The
subsurface has the following characteristics:

- Porosity $= 0.35$
- Organic content $= 0.03$
- Degree of water saturation $= 45\%$

- Subsurface temperature = 25°C
- Dry bulk density of soil = 1.6 g/cm³
- Total bulk density of soil = 1.8 g/cm³

Estimate the extracted soil vapor concentration at the start of the soil-venting project.

Solution:

(a) From Table 2.5, the following physicochemical properties of benzene were obtained:

$$Molecular\ weight = 78.1$$
$$H = 5.55\ atm/M$$
$$P^{vap} = 95.2\ mm\text{-}Hg$$
$$\log K_{ow} = 2.13$$

(b) Use Table 2.4 to convert Henry's constant to a dimensionless value:

$$H^* = H/RT = (5.55)/[(0.082)(273 + 25)] = 0.23\ (dimensionless)$$

Use Equation (2.28) to find K_{oc},

$$K_{oc} = 0.63K_{ow} = 0.63\ (10^{2.13}) = (0.63)(135) = 85$$

Use Equation (2.26) to find K_p,

$$K_p = f_{oc}K_{oc} = (0.03)\ (85) = 2.6\ L/kg$$

(c) Use Equation (2.40) to estimate the soil vapor concentration of benzene in equilibrium with this benzene concentration in soil:

$$X = \left\{ \frac{\left[\frac{(\phi_w)}{H} + \frac{(\rho_b)K_p}{H} + (\phi_a) \right]}{\rho_t} \right\} \times G$$

$$500 = \left\{ \frac{\left[\frac{(0.35)(0.45)}{0.23} + \frac{(1.6)(2.6)}{0.23} + (0.35)(1 - 45\%) \right]}{1.8} \right\} \times G$$

So, $G = 47.5\ mg/L = 47,500\ mg/m^3$

(d) Convert the vapor concentration into a volume concentration by using Equation (2.1):

$$1 \text{ ppmV benzene} = (78.1)/24.5 = 3.2 \text{ mg/m}^3 \text{ @25°C}$$
$$47{,}500 \text{ mg/m}^3 = 47{,}500 \div 3.2 = 14{,}800 \text{ ppmV}$$

Discussion:

The actual concentration of the extracted vapor would be smaller than 14,800 ppmV, due to the fact that not all the air flows through the impacted zone. In addition, limitations of mass transfer were not considered in these calculations.

Example 5.6: Estimate the Extracted Vapor Concentration (in the Absence of the Free Product)

For the site discussed in Example 5.5, the average benzene concentration of the soil samples, taken from the impacted zone, dropped to 250 mg/kg after three months of soil venting. Estimate the extracted soil vapor concentration from soil venting.

Solution:

Since the equilibrium constants (i.e., K_p and H) will stay the same, and assuming the volumetric water content also stays the same, the extracted vapor concentration (G) will drop to half of the initial value because the soil concentration dropped to half of the initial value.

$$250 = \left\{ \frac{\left[\frac{(0.35)(0.45)}{0.23} + \frac{(1.6)(2.6)}{0.23} + (0.35)(1-45\%) \right]}{1.8} \right\} \times G$$

So, $G = 47.5 \times (250/500) = 23.75 \text{ mg/L} = 23{,}750 \text{ mg/m}^3 = 7.400 \text{ ppmV}$

Discussion:

1. The equilibrium relationships used in this book are linear (i.e., $G = HC$ and $S = K_p C$), so that the relationship between X and G is also linear.

2. Mass-transfer limitations may play a more important role when the soil concentration becomes lower, which may make the extracted vapor concentration smaller than the calculated value.

5.2.3 Radius of Influence and Pressure Profile

Selecting the number and locations of vapor extraction wells is one of the major tasks in design of *in situ* soil vapor extraction systems. The decisions are typically based on the radius of influence (R_I), which can be defined as the distance from the extraction well where the pressure drawdown is very small (P @ $R_I \approx 1$ atm). The most accurate and site-specific R_I values should be determined from pilot testing. The pressure drawdown data at the extraction well and the observation wells can be plotted as a function of the radial distance from the extraction well on a semilog plot to determine the R_I of the extraction well. The approach is similar to the distance–drawdown method for aquifer tests, as described in Section 2.4.3. The R_I is commonly chosen to be the distance where the pressure drawdown is less than 1% of the vacuum in the extraction well.

The field test data can also be analyzed by using the flow equation, which describes the subsurface air flow. The subsurface is usually heterogeneous, and the air flow through it can be very complex. As a simplified approximation, a flow equation was derived for a fully confined radial gas flow system in a permeable formation having uniform and constant properties [1–4]. These references are the basis for much of the discussion on soil venting in this chapter.

For the steady-state radial flow subject to the boundary conditions ($P = P_w$ @ $r = R_w$ and $P = P_{atm}$ @ $r = R_I$), the pressure distribution in the subsurface can be derived as:

$$P_r^2 - P_w^2 = \left(P_{RI}^2 - P_w^2\right)\frac{\ln(r/R_w)}{\ln(R_I/R_w)} \tag{5.3}$$

where

P_r = pressure at a radial distance r from the vapor extraction well
P_w = pressure at the vapor extraction well
P_{RI} = pressure at the radius of influence (= atmospheric pressure or a preset value)
r = radial distance from the vapor extraction well
R_I = radius of influence where pressure is equal to the atmospheric pressure or a preset value
R_w = well radius of the vapor extraction well

Equation (5.3) can be used to determine the R_I of a vapor extraction well, if the pressure drawdown data of the extraction well and that of a monitoring well (or drawdown data of two monitoring wells) are known. As shown in Equation (5.3), the flow rate and the permeability of the formation are not included in this equation. The R_I can also be estimated from the vapor extraction rate and the pressure drawdown data in the extraction well (see Section 5.2.4).

If no pilot tests are conducted, an estimate could be made based on previous experiences. The R_I values ranging from 30 ft (9 m) to 100 ft (30 m) are reported in the literature, and typical pressures in the extraction wells range from 0.90 to 0.95 atm [3]. Shallower wells, less permeable subsurface, and lower applied vacuum in the extraction well generally correspond to smaller R_I values.

Example 5.7: Determine the Radius of Influence of a Soil-Venting Well by Using the Pressure Drawdown Data (Pressure Data Are Given in the Atmospheric Unit)

Determine the radius of influence of a soil-venting well using the following information:

- Pressure at the extraction well = 0.9 atm
- Pressure at a monitoring well 30 ft away from the venting well = 0.98 atm
- Diameter of the venting well = 4 in.

Solution:

(a) Let us define the R_I as the location where P is equal to the atmospheric pressure. The R_I can be found by using Equation (5.3) as:

$$P_r^2 - P_w^2 = \left(P_{RI}^2 - P_w^2\right)\frac{\ln(r/R_w)}{\ln(R_I/R_w)}$$

$$(0.98)^2 - (0.9)^2 = (1.0^2 - 0.9^2)\frac{\ln[30/(2/12)]}{\ln[R_I/(2/12)]}$$

$R_I = 118$ ft

(b) For comparison, let us now define the R_I as the location where the drawdown is equal to 1% of the vacuum in the extraction well:

The vacuum in the extraction well = 1 − 0.9 = 0.1 atm

Thus, $P_{RI} = 1 − (0.1)(1\%) = 0.999$ atm

$$(0.98)^2 - (0.9)^2 = (0.999^2 - 0.9^2)\frac{\ln[30/(2/12)]}{\ln[R_I/(2/12)]}$$

$R_I = 110$ ft

Discussion:

The R_I value from part (b), 110 ft, is about 7% smaller than that from part (a).

Example 5.8: Determine the Radius of Influence of a Soil-Venting Well by Using the Pressure Drawdown Data (Pressure Data Are Given in Inches of Water)

Determine the radius of influence of a soil-venting well using the following information:

- Pressure at the extraction well = 48-in. water vacuum
- Pressure at a monitoring well 40 ft away from the extraction well = 8-in. water vacuum
- Diameter of the vapor extraction well = 4 in.

Strategy:

The pressure data are expressed in inches of water. We need to convert them to the atmospheric unit or convert the atmospheric pressure to inches of water. A pressure of 1 atmosphere is equivalent to 33.9 ft of water column.

Solution:

(a) Pressure at the extraction well = 48-in. water vacuum

$$= 33.9 - (48/12) = 29.9 \text{ ft of water} = (29.9/33.9) = 0.88 \text{ atm}$$

Pressure at the monitoring well = 8-in. water vacuum

$$= 33.9 - (8/12) = 33.23 \text{ ft of water} = (33.23/33.9) = 0.98 \text{ atm}$$

(b) Let us define the R_I as the location where P is equal to the atmospheric pressure. The R_I can be found by using Equation (5.3) as:

$$(0.98)^2 - (0.88)^2 = (1.0^2 - 0.88^2) \frac{\ln[40/(2/12)]}{\ln[R_I/(2/12)]}$$

$R_I = 128 \text{ ft}$

(c) For comparison, let us now define the R_I as the location where the drawdown is equal to 1% of the vacuum in the extraction well:

Thus, $P_{RI} = 1 - (1 - 0.88)(1\%) = 0.9988 \text{ atm}$

$$(0.98)^2 - (0.88)^2 = (0.9988^2 - 0.88^2) \frac{\ln[40/(2/12)]}{\ln[R_I/(2/12)]}$$

$R_I = 120 \text{ ft}$

Discussion:

The R_I value from part (c), 120 ft, is about 7% smaller than that from part (b).

Example 5.9: Estimate the Pressure Drawdown in a Soil-Venting Monitoring Well

Using the pressure drawdown data given in Example 5.8, estimate the pressure drawdown (vacuum) in a monitoring well that is 20 ft away from the extraction well.

Strategy:

Example 5.8 gives the pressure drawdown data at (1) the monitoring well ($P = 0.88$ atm), (2) 40 ft away from the monitoring well ($P = 0.98$ atm), and (3) the R_I ($P = 1$ atm). We can use any two of these three to estimate the pressure drawdown in a well that is 20 ft away from the extraction well.

Solution:

(a) First, use the data of the extraction well and the monitoring well ($r = 40$ ft). The pressure at the monitoring well ($r = 20$ ft) can be found by using Equation (5.3) as:

$$P_r^2 - (0.88)^2 = (0.98^2 - 0.88^2)\frac{\ln[20/(2/12)]}{\ln[40/(2/12)]}$$

$P_r = 0.968$ atm $= 13.0$ in. of water (vacuum)

(b) We can also use the data of the extraction well and the R_I. The pressure at the monitoring well ($r = 20$ ft) can be found by using Equation (5.3) as:

$$P_r^2 - (0.88)^2 = (1.0^2 - 0.88^2)\frac{\ln[20/(2/12)]}{\ln[128/(2/12)]}$$

$P_r = 0.968$ atm $= 13.0$ in. of water (vacuum)

(c) We can also use the data of the monitoring well ($r = 40$ ft) and the R_I. The pressure at the monitoring well ($r = 20$ ft) can be found by using Equation (5.3) as:

$$P_r^2 - (0.98)^2 = (1.0^2 - 0.98^2)\frac{\ln[20/40]}{\ln[128/40]}$$

$P_r = 0.968$ atm $= 13.0$ in. of water (vacuum)

Discussion:

All three approaches yield the same result.

5.2.4 Vapor Flow Rates

The radial Darcy velocity, u_r, in homogeneous soil systems can be expressed as [2]:

$$u_r = \left(\frac{k}{2\mu}\right) \frac{\left[-\frac{P_w}{r \ln(R_w/R_I)}\right]\left[1-\left(\frac{P_{RI}}{P_w}\right)^2\right]}{\left\langle 1+\left[1-\left(\frac{P_{RI}}{P_w}\right)^2\right]\frac{\ln(r/R_w)}{\ln(R_w/R_I)}\right\rangle^{0.5}} \tag{5.4}$$

where u_r is the vapor flow velocity at a radial distance r away from the extraction well. The velocity at the wellbore, u_w, can be found by replacing r with R_w in Equation (5.4):

$$u_w = \left(\frac{k}{2\mu}\right)\left[\frac{P_w}{R_w \ln(R_w/R_I)}\right]\left[1-\left(\frac{P_{RI}}{P_w}\right)^2\right] \tag{5.5}$$

The volumetric vapor flow rate entering the extraction well, Q_w, can then be found as:

$$Q_w = (2\pi R_w H)(u_w)$$

$$= H\left(\frac{\pi k}{\mu}\right)\left[\frac{P_w}{\ln(R_w/R_I)}\right]\left[1-\left(\frac{P_{RI}}{P_w}\right)^2\right] \tag{5.6}$$

where H is the perforation interval of the extraction well.

To convert the vapor flow rate entering the well to the flow rate discharged to the atmosphere (Q_{atm}), where $P = P_{atm} = 1$ atm, the following relationship can be used:

$$Q_{atm} = \left(\frac{P_{well}}{P_{atm}}\right)Q_{well} \tag{5.7}$$

Example 5.10: Estimate the Extracted Vapor Flow Rate of a Soil-Venting Well

A soil-venting well (4-in. diameter) was installed at a site. The pressure in the extraction well is 0.9 atm and the radius of influence of this soil-venting well has been determined to be 50 ft.

Calculate the steady-state flow rate entering the well per unit well screen length, vapor flow rate in the well, and the vapor rate at the extraction pump discharge by using the following additional information:

- Permeability of the formation = 1 darcy
- Well screen length = 20 ft
- Viscosity of air = 0.018 cP
- Temperature of the formation = 20°C

Strategy:

We need to perform a few unit conversions first:

   1 atm = 1.013×10^5 N/m²

   1 darcy = 10^{-8} cm² = 10^{-12} m²

   1 poise = 100 centipoise = 0.1 N/s/m²

   (So, 0.018 centipoise = 1.8×10^{-4} poise = 1.8×10^{-5} N/s/m²)

Solution:

(a) The velocity at the wellbore, u_w, can be found by using Equation (5.5):

$$u_w = \left(\frac{k}{2\mu}\right)\left[\frac{P_w}{R_w \ln(R_w/R_I)}\right]\left[1-\left(\frac{P_{RI}}{P_w}\right)^2\right]$$

$$= \left[\frac{1\times10^{-12}}{2(1.8\times10^{-5})}\right]\left[\frac{(0.9)(1.013\times10^5)}{[(2/12)(0.3048)] \times \ln[(2/12)/50]}\right]\left[1-\left(\frac{1}{0.9}\right)^2\right]$$

$$= (2.78\times10^{-8})(-3.15\times10^5)(-0.2346)$$

$$= 2.05\times10^{-3}\,\text{m/s} = 0.123\ \text{m/min} = 177\ \text{m/day}$$

(b) The vapor flow rate entering the well per unit screen interval can be found by using Equation (5.6):

$$\frac{Q_w}{H} = 2\pi R_w u_w$$

$$= 2\pi[(2/12)(0.3048)\ \text{m}](0.123\ \text{m/min}) = 0.039\ \text{m}^3/\text{min/m}$$

(c) The vapor flow rate in the well = $(Q_w/H) \times H = (0.039\ \text{m}^3/\text{min/m})$
$[(20\ \text{ft})(0.3048\ \text{m/ft})]$
$= 0.24\ \text{m}^3/\text{min} = 8.4\ \text{ft}^3/\text{min}$

(d) The vapor flow rate at the exhaust of the extraction pump can be calculated from Equation (5.7):

$$Q_{atm} = \left(\frac{P_{well}}{P_{atm}}\right) Q_{well} = \left(\frac{0.9}{1}\right)(0.24)$$

$$= 0.216 \text{ m}^3/\text{min} = 7.6 \text{ ft}^3/\text{min}$$

Discussion:

Using consistent units in Equation (5.6) is very important. In the calculations shown here, the pressure is expressed in N/m^2, the distance in m, the permeability in m^2, and the viscosity in $N/s/m^2$. Consequently, the calculated velocity is in m/s.

Example 5.11: Estimate the Extracted Vapor Flow Rate of a Soil-Venting Well

A soil-venting well (4-in. diameter) was installed at a site. The pressure in the extraction well is 0.9 atm, and the radius of influence of this soil-venting well has been determined to be 50 ft. From Example 5.10, the radial Darcy velocity right outside the well casing was determined as 177 m/day. Calculate the radial Darcy velocity at 20 feet away from the center of the venting well by using Equation (5.4).

Solution:

(a) The radial air flow velocity at 20 ft away from the extraction well can be found by using Equation (5.4):

$$u_r = \left[\frac{10^{-12}}{2(1.8\times10^{-5})}\right] \frac{\left[\frac{(0.9)(1.013\times10^5)}{[(20)(0.3048)]\times \ln[(2/12)/50]}\right]\left[1-\left(\frac{1}{0.9}\right)^2\right]}{\left\langle 1+\left[1-\left(\frac{1}{0.9}\right)^2\right]\frac{\ln[20/(2/12)]}{\ln[(2/12)/50]}\right\rangle^{0.5}}$$

$$= (2.78\times10^{-8})(-2.64\times10^3)(-0.2346) \div (1.197)^{0.5}$$

$$= 1.574 \times 10^{-5} \text{ m/s} = 9.44\times10^{-4} \text{ m/min} = 1.36 \text{ m/day}$$

(b) For comparison, the radial air flow velocity can also be found as $Q = (2\pi r_1 H)v_1 = (2\pi r_2 H)v_2$ (assuming that the gas is incompressible and one-dimensional radial flow):

$$\text{Thus } r_1 v_1 = r_2 v_2$$

$$(2/12 \text{ ft})(177 \text{ m/day}) = (20)(v_2)$$

$$v_2 = \text{radial Darcy velocity 20 ft away}$$
$$= 1.48 \text{ m/day}$$

Discussion:

1. The answers from parts (a) and (b) should be the same. The apparent difference is from truncation errors.

2. The Darcy velocity at 20 feet away from the well is relatively slow, at 1.4 m/day.

Example 5.12: Estimate the Radius of Influence of a Soil-Venting Well by Using the Extracted Vapor Flow Rate

Determine the radius of influence of a soil-venting well using the following information:

- Pressure at the venting well = 0.85 atm
- Flow rate measured at the extraction pump discharge = 0.21 m³/min
- Well screen length = 4 m
- Diameter of the venting well = 0.1 m
- Permeability of the formation = 1.0 darcy
- Viscosity of air = 1.8×10^{-4} poise
- Temperature of the formation = 20°C

Strategy:

This problem can be viewed as the reverse of Example 5.10, in which the radius of influence was given for estimation of the vapor extraction flow rate. In this problem, the flow rate was given to estimate the radius of influence. As in the previous example, a few unit conversions need to be performed first:

- 1 atm = 1.013×10^5 N/m²
- 1 darcy = 10^{-8} cm² = 10^{-12} m²
- 1 poise = 100 centipoise = 0.1 N/s/m²
- (So, 0.018 centipoise = 1.8×10^{-4} poise = 1.8×10^{-5} N/s/m²)

Solution:

(a) The vapor flow rate entering the extraction well can be found by using Equation (5.7):

$$Q_{atm} = \left(\frac{P_{well}}{P_{atm}} \right) Q_{well} = 0.21 = \left(\frac{0.85}{1} \right) Q_{well}$$

$$Q_{well} = 0.24 \text{ m}^3/\text{min} = 0.004 \text{ m}^3/\text{s}$$

(b)   The radius of influence can be found by using Equation (5.6):

$$\frac{Q_w}{H} = \frac{0.004}{4} = 0.001$$

$$= \left(\frac{\pi k}{\mu}\right)\left[\frac{P_w}{\ln(R_w/R_I)}\right]\left[1-\left(\frac{P_{atm}}{P_w}\right)^2\right]$$

$$= \left(\frac{\pi(10^{-12})}{1.8\times10^{-5}}\right)\left[\frac{(0.85)(1.013\times10^5)}{\ln(0.05/R_I)}\right]\left[1-\left(\frac{1}{0.85}\right)^2\right]$$

$$R_I = 16.04 \text{ m}$$

Discussion:

Using consistent units is critical for successful calculations in this example. Specifically, the flow rate is given in m^3/min, but it needs to be converted to m^3/s to match the viscosity units in Equation (5.6).

Example 5.13: Estimate the Time to Flush One Pore Volume

A soil-venting well was installed in the middle of a plume, and the vapor extraction rate was found to be 20 ft^3/m. Assume that the well created a perfect radial flow with a radius of influence of 50 ft and a thickness of 20 ft. Find the time needed to flush one pore volume of the capture zone. The porosity of the subsurface is 0.4, and the volumetric water content is 0.15.

Solution:

(a)   The volume of the zone captured by this extraction well $= \pi(R_I)^2 H$
    $= (\pi) \times (50)^2 \times (20) = 157,100 \text{ ft}^3$

(b)   The volume of the air void = (volume of the soil)(ϕ_a) = (volume of the soil)($\phi - \phi_w$)
    $= (157,100)(0.4 - 0.15) = 39,270 \text{ ft}^3$

(c)   Time required to flush one pore volume = (void volume) ÷ (air flow rate)
    $= (39,270) \div 20 = 1,960 \text{ min}$

Discussion:

1. The flow rate of 20 ft^3/m is the measured flow rate at the surface. The actual flow rate in the subsurface should be slightly higher because it is under vacuum.

2. This example assumed a perfect ideal flow, but it would not happen in reality. In other words, the time required to flush one pore volume of the impacted zone will be much longer than 1,960 min.

5.2.5 COC Removal Rate

The COC removal rate ($R_{removal}$) can be estimated by multiplying the extracted vapor flow rate (Q) with the vapor concentration (G):

$$R_{removal} = (G)(Q) \qquad (5.8)$$

Care should be taken to have G and Q in consistent units, and G should be in mass-concentration units. Equation (5.1) can be used to estimate the initial vapor concentration if the free-product phase is present, while the procedure as illustrated in Example 5.5 can be used to estimate the extracted vapor concentration in the absence of the free-product phase. It is worthwhile to note again that the calculated vapor concentrations are the ideal and equilibrium values. The actual values should only be fractions of these values, mainly due to the fact that not the entire air stream passes through the impacted zone and that limitations of mass transfer exist (the system will not reach equilibrium in most, if not all, cases). Nevertheless, the calculated values provide useful information. One can compare them with the actual data from the collected samples and establish the correlation between them. The calculated data can then be calibrated, adjusted, and used for later predictions.

For example, if we know that only a fraction η of the air flows through the impacted zone, Equation (5.8) should be modified as:

$$R_{removal} = [(\eta)(G)](Q) \qquad (5.9)$$

The removal rate estimated from Equation (5.9) still represents the upper limit of the vapor concentration because it does not consider mass-transfer limitations. The factor η can be considered as an overall efficiency factor if it takes into account the percentage of flow through the impacted zone and the limitations of mass transfer.

The following procedure can be used to determine the COC removal rate:

Step 1: Determine the extraction vapor flow rate from field measurements or from the procedure described in Section 5.2.4.

Step 2: Estimate the extracted vapor concentration using Equation (5.1) if the free-product phase is present, while the procedure illustrated in Example 5.5 can be used to estimate the extracted vapor concentration in the absence of the free-product phase.

Step 3: Convert the vapor concentration into a mass concentration by using Equation (2.1).

Step 4: Adjust the calculated concentration from Step 3 by an overall efficiency factor, η.

Step 5: Calculate the mass removal rate by multiplying the flow rate (from Step 1) with the adjusted concentration (from Step 4).

Information needed for this calculation

- Extracted vapor flow rate, Q
- Extracted vapor concentration, G
- Overall efficiency factor relative to the theoretical removal rate, η

Example 5.14: Estimate the COC Removal Rate (in the Presence of Free-Product Phase)

Recently, a gasoline spill occurred at a gasoline station and caused subsurface contamination. A soil-venting well (4-in. diameter) was installed at a site. The pressure in the extraction well is 0.9 atm, and the radius of influence of this soil-venting well has been determined to be 50 ft.

Estimate the COC removal rate at the beginning of the project using the following information:

- Permeability of the formation = 1 darcy
- Well screen length = 20 ft
- Viscosity of air = 0.018 centipoise
- Temperature of the formation = 20°C

Solution:

(a) The flow rate has been determined in Example 5.10 to be 0.216 m^3/min, or 7.6 ft^3/min.

(b) Assuming the free-product phase is present, the saturated vapor concentration corresponding to the fresh gasoline is 1,340,000 ppmV, or 1,343 g/m^3 (see Example 5.1). On the other hand, the saturated vapor concentration corresponding to the weathered gasoline is 49,000 ppmV, or 226 g/m^3.

(c) Assuming the overall efficiency factor is equal to unity, the removal rate can be found from Equation (5.9) as:

$$R_{removal} = [(\eta)(G)](Q)$$
$$= [(1.0)(1{,}343 \text{ g/m}^3)](0.216 \text{ m}^3/\text{min})$$
$$= 290 \text{ g/min} = 0.64 \text{ lb/min} = 920 \text{ lb/day (for the fresh gasoline)}$$
$$= [(1.0)(226 \text{ g/m}^3)](0.216 \text{ m}^3/\text{min})$$
$$= 48.8 \text{ g/min} = 0.107 \text{ lb/min} = 155 \text{ lb/day (for the weathered gasoline)}$$

Discussion:

1. The extracted vapor flow rate in this example is relatively small, at 7.6 ft³/min. However, the calculated removal rates, 920 lb/day for the fresh gasoline and 155 lb/day for the weathered gasoline, are extraordinarily high. If the removal rate can be sustained at this level, the site would be cleaned up in a matter of days. Unfortunately, this is not the case. It normally takes months, if not longer, for a typical soil-venting project to reach completion. The overall efficiency factor was set as unity, which is extraordinarily high.

2. Since gasoline is a mixture of compounds, the removal rate will drop, as the more volatile ones have left the formation (as indicated by the five times lower removal rate of the weathered gasoline). However, the value of 155 lb/day corresponding to the weathered gasoline is still on the high side because the limitations of mass transfer were not included in this calculation. The removal rate should drop further after the free-product phase disappears.

Example 5.15: Estimate the COC Removal Rate (in the Absence of the Free-Product Phase)

A subsurface is impacted by benzene. The average benzene concentration of the soil samples, taken from the impacted zone, was 500 mg/kg. A soil-venting well (4-in. diameter) was installed at a site. The pressure in the extraction well is 0.9 atm, and the radius of influence of this soil-venting well has been determined to be 50 ft.

Estimate the benzene removal rate at the beginning of the project using the following information:

- Permeability of the formation = 1 darcy
- Well screen length = 20 ft
- Viscosity of air = 0.018 centipoise
- Temperature of the formation = 20°C
- Porosity = 0.35
- Organic content = 0.03
- Water saturation = 45%
- Dry bulk density of soil = 1.6 g/cm³
- Total bulk density of soil = 1.8 g/cm³

Solution:

(a) The flow rate has been determined in Example 5.10 to be 0.216 m³/min, or 7.6 ft³/min.

(b) The subsurface data are the same as those in Example 5.5, and the extracted benzene vapor concentration has been determined to be 47.5 mg/L, or 47.5 g/m³.

(c) Assuming the overall efficiency factor is equal to 1, the removal rate can be found from Equation (5.9) as:

$$R_{removal} = [(\eta)(G)](Q)$$
$$= [(1.0)(47.5 \text{ g/m}^3)](0.216 \text{ m}^3/\text{min})$$
$$= 10.26 \text{ g/min} = 14{,}770 \text{ g/day} = 32.5 \text{ lb/day}$$

Discussion:

The estimated value of 32.5 lb/day is on the high side because the overall efficiency factor is assumed to be unity. In addition, the removal rate would drop because the benzene concentration in the subsurface decreases as the venting project progresses.

5.2.6 Cleanup Time

Once the COC removal rate is determined, the cleanup time ($T_{cleanup}$) can be estimated as:

$$T_{cleanup} = M_{spill}/R_{removal} \tag{5.10}$$

where M_{spill} is the amount of spill to be removed. M_{spill} can be found by using Equation (5.11):

$$M_{spill} = (X_{initial} - X_{cleanup})(M_s) = (X - X_{cleanup})[(V_s)(\rho_t)] \tag{5.11}$$

where $X_{initial}$ is the average initial COC concentration in soil, $X_{cleanup}$ is the soil cleanup level, M_s is the mass of the impacted soil, V_s is the volume of the impacted soil, and ρ_t is the total bulk density of the soil. If the cleanup level is very low compared to the initial COC concentration, it can be deleted from Equation (5.11) as a factor of safety for design.

These two equations appear simple. However, estimation of the cleanup time is complicated by the fact that the COC removal rate is changing. The rate decreases as the amount of the COCs left in the soil decreases. One approach is to divide the cleanup into several time intervals. The removal rate for each interval is determined and used to estimate the cleanup time for each interval. The total cleanup time can then be derived from summing the cleanup time of each interval. The following steps detail this approach:

Step 1: Determine the maximum possible COC concentration in soil in the absence of free product, X_{sat} (see Example 5.3). If the average concentration of the soil samples exceeds this value, the free-product phase is present. Go to Step 2. If the average concentration of the samples is smaller, the free-product phase is absent. Go to Step 5.

Step 2: Estimate the extracted vapor concentration using Equation (5.1) and then calculate the mass removal rate using Equation (5.9).

Step 3: Determine the amount of COCs to be removed before the disappearance of the free-product phase by using modified Equation (5.11) as

$$M_{removal} = (X_{initial} - X_{sat})(M_s) = (X_{initial} - X_{sat})[(V_s)(\rho_t)] \qquad (5.12)$$

Step 4: Determine the required time for removal of the free product by using data from Steps 2 and 3 and Equation (5.10).

Step 5: Divide the $(X_{sat} - X_{cleanup})$ value into a few intervals. Use the average X of each interval to estimate the vapor concentration (see Example 5.5), and then calculate the mass removal rate using Equation (5.9). If no free-product phase is present initially, replace X_{sat} with $X_{initial}$ in this step.

Step 6: Determine the amount of COCs to be removed in each interval by using modified Equation (5.11):

$$M_{removal} = (X_{initial} - X_{final})(M_s) = (X_{initial} - X_{final})[(V_s)(\rho_t)] \qquad (5.13)$$

Step 7: Determine the required cleanup time for each interval by using data from Steps 5 and 6 and Equation (5.10).

Step 8: Sum the required time for each interval to calculate the total cleanup time.

Information needed for this calculation:

- COC concentrations of soil samples
- Henry's constant of the COC
- Organic–water partition coefficient, K_{ow}
- Organic content, f_{oc}
- Porosity, ϕ
- Degree of water saturation

- Dry bulk density of soil, ρ_b
- Total bulk density of soil, ρ_t

Example 5.16: Estimate the Cleanup Time (in the Presence of Free-Product Phase)

Recently, a gasoline spill occurred at a gasoline station and caused subsurface contamination. A soil-venting well (4-in. diameter) was installed at a site. The pressure in the extraction well is 0.9 atm, and the radius of influence of this soil-venting well has been determined to be 50 ft.

Estimate the required cleanup time for total petroleum hydrocarbon (TPH) using the following information:

- Permeability of the formation = 1 darcy
- Well screen length = 20 ft
- Viscosity of air = 0.018 centipoise
- Temperature of the formation = 20°C
- Porosity = 0.35
- Organic content in soil = 0.01
- Degree of water saturation = 40%
- Dry bulk density of soil = 1.6 g/cm³
- Total bulk density of soil = 1.8 g/cm³
- Size of the plume = 6,500 ft³
- Initial average TPH concentration in soil = 6,000 mg/L
- Required cleanup level = 100 mg/L
- Overall efficiency factor relative to theoretical removal rate = 0.11

Solution:

(a) The flow rate has been determined in Example 5.10 to be 0.216 m³/min, or 7.6 ft³/min.

Presence of the Free-Product Phase

(b) Determine the maximum possible TPH concentration in soil absent of free product, X_{sat} (use the procedure illustrated in Example 5.3):

Since no Henry's constant and K_{ow} data are available for gasoline, we use those of toluene, one of the common gasoline components, as an approximation:

Use Table 2.4 to convert the Henry's constant to a dimensionless value:

$$H^* = H/RT = (6.7)/[(0.082)(273 + 20)] = 0.28 \text{ (dimensionless)}$$

Use Equation (2.28) to find K_{oc}:

$$K_{oc} = 0.63K_{ow} = 0.63 \ (10^{2.73}) = (0.63)(537) = 338$$

Use Equation (2.26) to find K_p:

$$K_p = f_{oc}K_{oc} = (0.01) \ (338) = 3.4 \text{ L/kg}$$

Use the saturated gasoline vapor concentration of weathered gasoline, 226 mg/L (from Example 5.1) and Equation (5.2) to estimate X_{sat}:

$$X_{sat} = \left\{ \frac{\left[\frac{(0.35)(40\%)}{0.28} + \frac{(1.6)(3.4)}{0.28} + (0.35)(1 - 40\%) \right]}{1.8} \right\} \times (226)$$

$$= 2{,}528 \text{ mg/kg}$$

(c) Determine the amount of TPH to be removed before the disappearance of the free-product phase by using Equation (5.12):

$$M_s = (V_s)(\rho_t) = (6{,}500 \text{ ft}^3)[(1.8 \times 62.4 \text{ lb/ft}^3)]$$
$$= 730{,}100 \text{ lb} = 332{,}000 \text{ kg}$$
$$M_{removal} = (X_{initial} - X_{sat})(M_s)$$
$$= (6{,}000 - 2{,}528 \text{ mg/kg})(332{,}000 \text{ kg})$$
$$= 1.153 \times 10^9 \text{ mg} = 1{,}153 \text{ kg}$$

(d) Estimate the extracted vapor concentration using Equation (5.1):

As determined in Example 5.1, the saturated gasoline vapor concentrations are 1,343 mg/L and 226 mg/L for the fresh and the weathered gasoline, respectively. Since the observed VOC concentrations of the extracted vapor often decrease exponentially over time, the geometric average of these two values is used as the average concentration for this interval:

$$G = \sqrt{(1{,}343)(226)} = 551 \text{ mg/L}$$

(e) Calculate the mass removal rate using Equation (5.9):

$$R_{removal} = [(\eta)(G)](Q)$$
$$= [(0.11)(551 \text{ g/m}^3)](0.216 \text{ m}^3/\text{min})$$
$$= 13.1 \text{ g/min} = 18.85 \text{ kg/day}$$

(f) Determine the required cleanup time by using data from (c) and (e) and Equation (5.10):

$$T_1 = M_{removal} \div R_{removal} = (1{,}153 \text{ kg}) \div 18.85 \text{ kg/day} = 61.2 \text{ day}$$

Absence of the Free-Product Phase

(g) At the end of the free-product removal, the TPH concentration in soil is 2,528 mg/kg, corresponding to a theoretical vapor concentration of 226 mg/L. The cleanup level of soil for this project is 100 mg/kg. The average of 2,528 and 100 is equal to 1,314. To estimate the required cleanup time, we divide it into two intervals. The first interval is the time required to reduce the concentration from 2,528 to 1,314 mg/kg, and the other is from 1,314 to 100 mg/kg.

(h) Determine the amount of TPH to be removed in the first interval by using Equation (5.13) as:

$$M_{removal} = (X_{initial} - X_{final})(M_s)$$
$$= (2{,}528 - 1{,}314 \text{ mg/kg})(332{,}000 \text{ kg})$$
$$= 4.03 \times 10^8 \text{ mg} = 403 \text{ kg}$$

For this interval, the initial theoretical vapor concentration is 226 mg/L (corresponding to 2,528 mg/kg), and the final theoretical vapor concentration (corresponding to 1,314 mg/kg) can be easily found as:

$$G_{final} = 226 \times (1{,}314/2{,}528) = 117 \text{ mg/L}$$

The geometric average of these two is used as the average concentration for this interval:

$$G = \sqrt{(226)(117)} = 163 \text{ mg/L}$$

Calculate the mass removal rate by using Equation (5.9):

$$R_{removal} = [(\eta)(G)](Q)$$
$$= [(0.11)(163 \text{ g/m}^3)](0.216 \text{ m}^3/\text{min})$$
$$= 3.87 \text{ g/min} = 5.58 \text{ kg/day}$$

Determine the required cleanup time by using Equation (5.10):

$$T_2 = M_{removal} \div R_{removal} = (403 \text{ kg}) \div 5.58 \text{ kg/day} = 72.2 \text{ day}$$

(i) In the second interval, the amount of the TPH mass to be removed is the same as that of the first interval, 403 kg. The initial theoretical vapor concentration is 117 mg/L (corresponding to 1,314 mg/kg), and the final theoretical vapor concentration (corresponding to 100 mg/kg) can be easily found as:

$$G_{final} = 117 \times (100/1,314) = 8.9 \text{ mg/L}$$

The geometric average of these two is used as the average concentration for this interval:

$$G = \sqrt{(117)(8.9)} = 32.3 \text{ mg/L}$$

Calculate the mass removal rate by using Equation (5.9):

$$\begin{aligned} R_{removal} &= [(\eta)(G)](Q) \\ &= [(0.11)(32.3 \text{ g/m}^3)](0.216 \text{ m}^3/\text{min}) \\ &= 0.77 \text{ g/min} = 1.1 \text{ kg/day} \end{aligned}$$

Determine the required cleanup time by using Equation (5.10):

$$T_3 = M_{removal} \div R_{removal} = (403 \text{ kg}) \div 1.1 \text{ kg/day} = 365 \text{ days}$$

The Entire Project

$$\begin{aligned} \text{The total cleanup time required} &= T_1 + T_2 + T_3 \\ &= 61.2 + 72.2 + 365 \\ &= 498 \text{ days} \end{aligned}$$

Discussion:

1. For the three intervals, the average mass removal rates drop significantly from 18.85 kg/day in the first interval to 5.6 kg/day, then to 1.1 kg/day in the third interval.

2. For the two intervals during the absence of free product, the second interval takes 365 days and the first interval takes only 72 days to remove the same amount of TPH.

3. The cleanup time of 498 days is not acceptable in most project applications. One may consider increasing the extraction flow rate or adding more wells.

4. Only two intervals were used to analyze the period between free-product disappearance and final cleanup. The estimate would be more accurate if more intervals were used.

5. If the free-product phase is not present initially, solve the problem by starting from part (g).

5.2.7 Effect of Temperature on Soil Venting

In a soil-venting project, the subsurface temperature will affect both the air flow rate and the vapor concentration. At a higher temperature, the vapor pressure of an organic compound would be higher. On the other hand, the higher subsurface temperature will yield a lower air flow rate because air viscosity increases with temperature:

$$\frac{\mu@T_1}{\mu@T_2} = \sqrt{\frac{T_1}{T_2}} \tag{5.14}$$

where T is the subsurface temperature, expressed in Kelvin or Rankine units. The ratio of the flow rates at two temperatures can be estimated by using Equation (5.15):

$$\frac{Q@T_1}{Q@T_2} = \sqrt{\frac{T_2}{T_1}} \tag{5.15}$$

As shown in Equation (5.15), the vapor flow rate will be lower at higher temperatures. However, since the vapor concentration will be much higher at a higher temperature, the mass removal rate will still be higher at a higher temperature.

Example 5.17: Estimate the Extracted Vapor Flow Rate of a Soil-Venting Well at Elevated Temperatures

A soil-venting well (4-in. diameter) was installed at a site. The pressure in the extraction well is 0.9 atm, and the radius of influence of this soil-venting well has been determined to be 50 ft.

If the subsurface temperature is raised to 30°C, what will be the vapor flow rate (all the other conditions being kept the same)? The extracted vapor flow rate has been estimated, as shown in Example 5.10, to be 7.6 ft³/min under the following conditions:

- Permeability of the formation = 1 darcy
- Well screen length = 20 ft
- Viscosity of air = 0.018 cP
- Temperature of the formation = 20°C

Solution:
The new air flow rate can be found by using Equation (5.15) as

$$\frac{Q@\,30^{\circ}C}{Q@\,20^{\circ}C} = \sqrt{\frac{273+20}{273+30}}$$

$$Q@\,30^{\circ}C = (7.6)(0.967) = 7.3 \text{ ft}^3/\text{min}$$

Discussion:
The temperature affects the air flow rate insignificantly. For a 10°C increase in temperature, the flow rate decreases by less than 4%.

5.2.8 Number of Vapor Extraction Wells

There are three main considerations in determining the number of vapor extraction wells necessary for a soil-venting project. First, a successful soil-venting project should have a sufficient number of extraction wells to cover the entire plume. In other words, the entire impacted zone should be within the influence of the wells, thus

$$N_{wells} = \frac{1.2(A_{plume})}{\pi R_I^2} \tag{5.16}$$

The factor of 1.2 is arbitrarily chosen to account for the overlapping of the influence areas among the wells as well as the fact that the peripheral wells may reach outside the impacted zone. Second, the number of wells should be sufficient to complete the site cleanup within an acceptable time frame.

$$R_{acceptable} = \frac{M_{spill}}{T_{cleanup,acceptable}} \tag{5.17}$$

$$N_{wells} = \frac{R_{acceptable}}{R_{removal}} \tag{5.18}$$

The minimum number of the vapor extraction wells should be the larger of the two that are determined from Equations (5.16) and (5.18).

The last, and probably the most important, consideration is the economical one. There is a trade-off between the number of wells and treatment cost. One can install more wells to shorten the cleanup time, but it may be more costly.

Example 5.18: Determine the Number of Venting Wells Required

For the soil-venting project described in Example 5.16, it is desired to clean up the site in 9 months. Determine the number of venting wells needed. The maximum cross-sectional area of the plume is equal to 9,000 ft².

Solution:

(a) The flow rate from one venting well has been determined to be 0.216 m³/min, or 7.6 ft³/min. At this flow rate, the cleanup will take 498 days (from Example 5.16). To meet the 9-month cleanup schedule, the removal rate should be 1.8 (= 498 ÷ 270) times faster. Therefore, we need to increase the flow rate by 1.8 times or to have two venting wells.

(b) The radius of influence of one venting well has been determined to be 50 ft. The number of wells needed to cover the plume can be determined by using Equation (5.16) as:

$$N_{wells} = \frac{1.2(A_{plume})}{\pi R_I^2} = \frac{1.2(9,000)}{\pi (50)^2} = 1.38$$

Therefore, two wells should be enough to cover the entire plume, unless the plume has a very long stripe shape.

(c) Based on these results, two venting wells would be required.

5.2.9 Sizing the Vacuum Pump (Blower)

The theoretical horsepower requirements ($hp_{theoretical}$) of vacuum pumps, blowers, or compressors for an ideal gas undergoing an isothermal compression (PV = constant) can be expressed as [5]:

$$hp_{theoretical} = 3.03 \times 10^{-5} P_1 Q_1 \ln \frac{P_2}{P_1} \qquad (5.19)$$

where
P_1 = intake pressure, lb_f/ft²
P_2 = final delivery pressure, lb_f/ft²
Q_1 = air flow rate at the intake condition, ft³/min

For an ideal gas undergoing an isentropic compression (PV^k = constant), the following equation is applicable for single-stage compressors [5]:

$$hp_{theoretical} = \frac{3.03 \times 10^{-5} k}{k-1} P_1 Q_1 \left[\left(\frac{P_2}{P_1} \right)^{(k-1)/k} - 1 \right] \qquad (5.20)$$

where k is the ratio of specific heat of gas at constant pressure to specific heat of gas at constant volume. For the typical soil-venting applications, it is appropriate to use $k = 1.4$.

For reciprocating compressors, the efficiencies (E) are generally in the range of 70% to 90% for isentropic and 50% to 70% for isothermal compression. The actual horsepower requirement can be found as:

$$hp_{actual} = \frac{hp_{theoretical}}{E} \qquad (5.21)$$

Example 5.19: Determine the Required Horsepower of the Vacuum Pump in Soil Venting

Two vapor extraction wells are installed. The design flow rate of each well is 40 ft³/min, and the design wellhead pressure is 0.9 atm. A vacuum pump is to serve both wells. Estimate the required horsepower of the vacuum pump.

Solution:

(a) The pressure in the extraction well, P_1

   $= 0.9$ atm $= (0.9)(14.7) = 13.2$ psi $= (13.2)(144)$ psi $= 1,905$ lb/ft²

(b) Assuming isothermal expansion, use Equation (5.19) to determine the theoretical power requirement as:

$$hp_{theoretical} = (3.03 \times 10^{-5}) P_1 Q_1 \ln \frac{P_2}{P_1}$$

$$= (3.03 \times 10^{-5})[(14.7)(144)][(2)(40)] \ln \frac{(14.7)(144)}{(1905)}$$

$$= 0.54\,hp$$

Assuming an isothermal efficiency of 60%, the actual horsepower required is determined by using Equation (5.21) as:

$$hp_{actual} = \frac{hp_{theoretical}}{E} = \frac{0.54}{60\%} = 0.9\,hp$$

(c) Assuming isentropic expansion, use Equation (5.20) to determine the theoretical power requirement as:

$$hp_{theoretical} = \frac{(3.03\times10^{-5})k}{k-1}P_1Q_1\left[\left(\frac{P_2}{P_1}\right)^{(k-1)/k}-1\right]$$

$$= \frac{(3.03\times10^{-5})(1.4)}{1.4-1}[(14.7)(144)][(40)(2)]$$

$$\times\left[\left(\frac{(14.7)(144)}{1,905}\right)^{(1.4-1)/1.4}-1\right]=0.55\ hp$$

Assuming an isentropic efficiency of 80%, the actual horsepower required is determined by using Equation (5.21) as:

$$hp_{actual} = \frac{hp_{theoretical}}{E} = \frac{0.55}{80\%} = 0.7\ hp$$

Discussion:

In soil-venting applications, the difference between the inlet and final discharge pressures is relatively small. Consequently, the theoretical power requirements for isothermal and isentropic compression are very similar, as illustrated in this example.

5.3 Soil Washing/Solvent Extraction/Soil Flushing

5.3.1 Description of the Soil-Washing Process

The majority of the organic and inorganic COCs contained in soil is associated with fines (i.e., clay or silt) that have large specific surface areas. These fine particles, in turn, are attached to sand and gravel, which are much larger in size, by compaction and adhesion. In this section, three technologies (soil washing, solvent extraction, and soil flushing) are discussed. They use solvents to extract or separate COCs from the soil matrix.

Soil washing is a water-based washing process. The major removal mechanisms include desorption of COCs from the soil grains, consequent dissolution into the washing fluid, and/or suspension of the clay and silt particles with bound COCs into the washing fluid. The COCs are readily washed off from sand and gravel, which often account for a large portion of the soil matrix. Separation of the sand and gravel from the heavily impacted clay and silt particles greatly reduces the volume of impacted soil. Soil washing makes further treatment or disposal easier.

Various chemicals can be added to the aqueous solution to enhance desorption or dissolution of the COCs. For example, an acidic solution is often used to extract heavy metals from impacted soil. Addition of chelating agents can help the dissolution of heavy metals into the aqueous solution, while addition of surface-active agents can help the dissolution of organics. Solvent extraction is similar to soil washing, except that solvents rather than aqueous solutions are employed to extract organic COCs from soil. Commonly used solvents include alcohol, liquefied propane and butane, and supercritical fluids.

Soil flushing differs from soil washing or solvent extraction in that it is an *in situ* process in which water or solvent flushes the impacted zone to desorb or dissolve the COCs. The elutriate is then collected from the wells or drains for further treatment.

5.3.2 Design of a Soil-Washing System

A mass-balance equation can be written to relate the COC concentrations in the soil before and after washing with the COC concentration in the spent washing fluid (assuming that the fresh washing fluid does not contain any COCs) as:

$$X_{initial} M_{s,wet} = S_{initial} M_{s,dry} + C_{initial} V_{soil\,moisture}$$
$$= S_{final} M_{s,dry} + C_{final} V_1 + C_{final} V_{soil\,moisture} \tag{5.22}$$

where

X_{inital} = measured COC concentration of the soil sample before washing (mg/kg)

$M_{s,wet}$ = mass of soil plus moisture before washing (kg)

$M_{s,dry}$ = mass of dry soil (kg)

$S_{initial}$ = initial COC concentration on the surface of the soil before washing (mg/kg)

S_{final} = final COC concentration on the surface of the washed soil (mg/kg)

$C_{initial}$ = COC concentration in the soil moisture before washing (mg/L)

C_{final} = COC concentration in the spent washing fluid (mg/L)

$V_{soil\,moisture}$ = volume of the soil moisture before washing (L)

V_1 = volume of the washing fluid used (L)

The term on the left-hand side of Equation (5.22) represents the total COC mass in soil before washing, which includes the mass adsorbed on the soil surface and that dissolved in the soil moisture (shown as the middle section of the equation). The terms in the last section of the equation represent the mass left on the soil surface and the mass dissolved in the liquid phase at the end of washing (the total volume of the liquid is equal to the sum of the washing fluid, V_l, and the soil moisture, $V_{moisture}$). Assuming the mass in the soil moisture is relatively small when compared to mass adsorbed on the soil surface before washing (i.e., $C_{initial}V_{soil\ moisture} \ll S_{initial}M_{s,dry}$) and the mass in soil moisture is relatively small compared to the sum of the mass adsorbed on the soil surface and that in the washing fluid after washing (i.e., $C_{final}V_{soil\ moisture} \ll (S_{final}M_{s,dry} + C_{final}V_l)$, Equation (5.22) can be simplified to

$$S_{initial}M_{s,dry} \approx S_{final}M_{s,dry} + C_{final}V_l \qquad (5.23)$$

These two assumptions are valid when the soil before washing is relatively dry and/or the COC is relatively hydrophobic.

If an equilibrium condition is achieved at the end of the washing, the COC concentration on the soil and that in the liquid can be related by the partition equation described in Chapter 2, Equation 2.24:

$$S_{final} = K_p C_{final} \qquad (5.24)$$

where K_p is the partition equilibrium constant. By inserting Equation (5.24) into Equation (5.23), the relationship between the initial and final COC concentrations on the soil surface can be expressed by Equation (5.25) or Equation (5.26) as:

$$\frac{S_{final}}{S_{initial}} = \frac{1}{1 + \left(\frac{V_l}{M_{s,dry}K_p}\right)} \qquad (5.25)$$

$$S_{final} = \frac{1}{1 + \left(\frac{V_l}{M_{s,dry}K_p}\right)} \times S_{initial} \qquad (5.26)$$

For a sequence of washers in series, the final COC concentration can be determined by:

$$\frac{S_{final}}{S_{initial}} = \frac{1}{1 + \left(\frac{V_{l,1}}{M_{s,dry}K_p}\right)} \times \frac{1}{1 + \left(\frac{V_{l,2}}{M_{s,dry}K_p}\right)} \times \frac{1}{1 + \left(\frac{V_{l,3}}{M_{s,dry}K_p}\right)} \times \cdots \qquad (5.27)$$

As shown in Example 2.38, the values of S (the adsorbed COC concentration on the soil surface) and X (the COC concentration of the soil sample) are relatively similar for compounds like benzene; while the ratio of the X and S values is essentially the ratio of the dry bulk density and total bulk density for very hydrophobic compounds such as pyrene. For either case, the following relationship is valid:

$$\frac{S_{final}}{S_{initial}} \approx \frac{X_{final}}{X_{initial}} \tag{5.28}$$

By inserting Equation (5.28) into Equations (5.25), (5.26), and (5.27), we can obtain:

$$\frac{X_{final}}{X_{initial}} \approx \frac{S_{final}}{S_{initial}} = \frac{1}{1+\left(\frac{V_l}{M_{s,dry}K_p}\right)} \tag{5.29}$$

$$X_{final} \approx \frac{1}{1+\left(\frac{V_l}{M_{s,dry}K_p}\right)} \times X_{initial} \tag{5.30}$$

$$\frac{X_{final}}{X_{initial}} \approx \frac{S_{final}}{S_{initial}} = \frac{1}{1+\left(\frac{V_{l,1}}{M_{s,dry}K_p}\right)} \times \frac{1}{1+\left(\frac{V_{l,2}}{M_{s,dry}K_p}\right)} \times \frac{1}{1+\left(\frac{V_{l,3}}{M_{s,dry}K_p}\right)} \times \cdots \tag{5.31}$$

The relationship among mass of soil before washing ($M_{s,wet}$), mass of dry soil ($M_{s,dry}$), dry bulk density (ρ_b), and total bulk density (ρ_t) can be found from the following linear relationship:

$$\frac{M_{s,dry}}{M_{s,wet}} \approx \frac{\rho_b}{\rho_t} \tag{5.32}$$

Example 5.20: Determine the Efficiency of Soil Washing

A sandy subsurface contains 500 mg/L of 1,2-dichloroethane (DCA) and 500 mg/L of pyrene. Soil washing is proposed to remediate the soil. A batch washer that can accommodate 1,000 kg of soil is designed. For each batch of operation, 1,000 gal of clean water is used as the washing fluid. Determine the final concentrations of these two COCs in the washed soil.
Use the following data from site assessment in design:

- Dry bulk density of soil = 1.6 g/cm³
- Total bulk density of soil = 1.8 g/cm³
- Fraction of organic carbon of aquifer materials = 0.005
- $K_{oc} = 0.63 K_{ow}$

Solution:

(a) From Table 2.5,

$$\log(K_{ow}) = 1.53 \text{ for 1,2-DCA} \rightarrow K_{ow} = 34$$
$$\log(K_{ow}) = 4.88 \text{ for pyrene} \rightarrow K_{ow} = 75{,}900$$

(b) Using the given relationship, $K_{oc} = 0.63 K_{ow}$, we obtain

$$K_{oc} = (0.63)(34) = 22 \text{ (for 1,2-DCA)}$$
$$K_{oc} = (0.63)(75{,}900) = 47{,}800 \text{ (for pyrene)}$$

(c) Using Equation (2.26), $K_p = f_{oc} K_{oc}$, and $f_{oc} = 0.005$, we obtain

$$K_p = (0.005)(22) = 0.11 \text{ L/kg (for 1,2-DCA)}$$
$$K_p = (0.005)(47{,}800) = 239 \text{ L/kg (for pyrene)}$$

(d) Use Equation (5.32) to find the mass of dry soil:

$$M_{s,dry} = (1{,}000) \times (1.6/1.8) = 889 \text{ kg}$$

Use Equation (5.30) to find the final concentration as (1,000 gal = 3,785 L):

$$X_{final} \approx \frac{1}{1 + \left(\frac{V_l}{M_{s,dry} K_p}\right)} \times X_{initial}$$

$$X_{final} \approx \frac{1}{1 + \left(\frac{3{,}785}{(889)(0.11)}\right)} \times 500 = 12.6 \text{ mg/L} \quad \text{for 1,2-DCA}$$

$$X_{final} \approx \frac{1}{1 + \left(\frac{3{,}785}{(889)(239)}\right)} \times 500 = 491 \text{ mg/L} \quad \text{for pyrene}$$

Discussion:

1. Pyrene is very hydrophobic, and its K_p value is very high. This example demonstrates that water washing is essentially ineffective in removing pyrene from soil. Addition of surfactants into the washing fluid, using organic solvents, or raising the temperature of the washing fluid should be considered.

2. The calculated values are based on an assumption that the liquid and the soil are in equilibrium. For a practical reactor design, an equilibrium condition is seldom reached. Consequently, the actual final concentration would be higher.

Example 5.21: Determine the Efficiency of Soil Washing
                    (Two Reactors in Series)

The single reactor described in Example 5.20 could not reduce the 1,2-DCA concentration to below 10 mg/L. An engineer proposed using two smaller washers in series. The washer still accommodates 1,000 kg of soil, but only 500 gal of fresh water is added to each washer. Can this system meet the cleanup requirements?

Solution:

Use Equation (5.31) to find the final concentration for two reactors in series as ($V_{1,1} = V_{1,2} = 500$ gal $= 1,893$ L):

$$X_{final} = \frac{1}{1+\left(\frac{V_{1,1}}{M_{s,dry}K_p}\right)} \times \frac{1}{1+\left(\frac{V_{1,2}}{M_{s,dry}K_p}\right)} \times X_{initial}$$

$$= \frac{1}{1+\left(\frac{1,893}{(889)(0.11)}\right)} \times \frac{1}{1+\left(\frac{1,893}{(889)(0.11)}\right)} \times 500 = 1.2 \text{ mg/L}$$

Discussion:

1. In both cases, the same amount of water, 1,000 gal, is used for 1,000 kg of soil. However, use of two small reactors in series yields a lower final concentration.

2. The calculated values are based on an assumption that the liquid and the soil are in equilibrium. For a practical reactor design, an equilibrium condition is seldom reached. Consequently, the actual final concentration would be higher.

5.4 Soil Bioremediation

5.4.1 Description of the Soil-Bioremediation Process

Soil bioremediation utilizes microorganisms or their metabolic products to degrade organic COCs in soil. Soil bioremediation can be conducted under aerobic or anaerobic conditions, but aerobic bioremediation is more popular. The final products of complete aerobic biodegradation of hydrocarbons are carbon dioxide and water.

Bioremediation may be conducted either *in situ* or *ex situ*. *Ex situ* soil-bioremediation processes are more developed and widely used than *in situ* processes. *Ex situ* bioremediation is typically performed using one of three systems: (1) static soil pile, (2) in-vessel, and (3) slurry bioreactor. The static soil pile is the most popular format. This approach treats soil stockpiled on the site with per-forated pipes embedded in the piles as the conduit for air supply. To minimize fugitive emission and potential secondary contamination from leachates, the stockpiles are usually covered on the top and lined at the bottom.

In situ treatment enhances the natural microbial activity of undisturbed soil in place to decompose organic COCs. A nutrient solution is often per-colated or injected into the subsurface to support the activities of the bio-degraders. Run-on and run-off controls and waste containment are often required. In a slurry bioreactor, impacted soil is mixed with a nutrient solu-tion under controlled operating conditions (i.e., optimal pH, temperature, dissolved oxygen, and mixing).

Microorganisms require moisture, oxygen (or absence of oxygen for anaerobic biodegradation), nutrients, and a suitable set of environmental factors to grow. The environmental factors include pH, temperature, and absence of toxic conditions. Table 5.2 summarizes the critical conditions for bioremediation.

5.4.2 Moisture Requirement

As shown in Table 5.2, the optimal moisture content for soil bioremediation is 25%–85% of the water-holding capacity. In most cases, soil moisture will be below or in the lower end of this range; therefore, addition of moisture is commonly needed.

The moisture present in the vadose zone is often quantified by a term called the *volumetric water content* or *degree of saturation*. Volumetric water content var-ies from zero to the value of porosity, while degree of saturation varies from zero to one and refers to the percentage of pore space occupied by moisture. For complete saturation, the volumetric water content is equal to porosity, and

TABLE 5.2

Critical Conditions for Bioremediation

Environmental Factor	Optimum Conditions
Available soil water	25%–85% water-holding capacity
Oxygen	Aerobic metabolism: >0.2 mg/L dissolved oxygen, air-filled pore space to be >10% by volume Anaerobic metabolism: oxygen concentration to be <1% by volume
Redox potential	Aerobes and facultative anaerobes: >50 millivolts Anaerobes: <50 millivolts
Nutrients	Sufficient N, P, and other nutrients (suggested C:N:P molar ratio of 120:10:1)
pH	5.5 to 8.5 (for most bacteria)
Temperature	15°C–45°C (for mesophiles)

Source: [6].

the degree of water saturation is 100%. The following formula can be used to determine the volume of water (V_{water}) needed for bioremediation.

$$V_{water} = \text{(volume of soil)} \times \text{(desired moisture content - initial moisture content)}$$
$$= (V_{soil})(\phi_{w,f} - \phi_{w,i})$$
$$= (V_{soil})[(\phi)(S_{w,f} - S_{w,i})] \tag{5.33}$$

where
   $\phi_{w,i}$ = initial volumetric moisture content
   $\phi_{w,f}$ = desired volumetric moisture content
   ϕ = porosity of soil
   $S_{w,i}$ = initial degree of saturation
   $S_{w,f}$ = desired degree of saturation

Example 5.22: Determine the Moisture Requirement for Soil Bioremediation

An underground storage tank (UST)-removal project resulted in 375 yd³ of soil impacted by gasoline that has to be treated before disposal. Bioremediation using static stockpiles has been selected as the treatment method. Determine the amount of water needed for the first spray.
   Use the following simplified assumptions in your calculation:

- Porosity of soil = 35%
- Initial degree of saturation = 20%

Solution:
   (a) Based on Table 5.2, the optimal volumetric water content for soil bioremediation is 25%–85% of the water-holding capacity. Without conducting an optimization study, the middle value of this range, 55%, is selected.
   (b) Water needed = (375)[(0.35)(55% − 20%)]
$$= 45.9 \text{ yd}^3 = 1{,}240 \text{ ft}^3 = 9{,}280 \text{ gal}$$

Discussion:
Addition of makeup water is often needed periodically.

5.4.3 Nutrient Requirements

Nutrients for microbial activity usually exist in the subsurface. However, with the elevated level of organic COCs, additional nutrients are often needed to support the bioremediation of the COCs. The nutrients to enhance microbial growth are assessed primarily on the nitrogen and phosphorus requirements.

As shown in Table 5.2, the suggested C:N:P ratio is 120:10:1. (Some other references suggest C:N:P = 100:10:1.) The ratio is on a molar basis. It means that every 120 moles of carbon requires 10 moles of nitrogen and 1 mole of phosphorus. For bioremediation, a feasibility study is always recommended. Determination of an optimal nutrient ratio should be part of the feasibility study. If no other information is available, the ratio mentioned here can be used. The example in this section will show that the amount of nutrients needed is relatively small, and so is the cost. Nutrients are often dissolved in water first, and this is then applied to the soil by spraying or irrigation.

To determine the nutrient requirements, the following procedure can be used:

Step 1: Determine the mass of the organics present in the impacted soil.

Step 2: Divide the mass of organics by its molecular weight to find the moles of the COCs.

Step 3: Multiply the moles of COCs from Step 2 by the number of Cs in the compound's formula.

Step 4: Determine the moles of nitrogen and phosphorus needed using the optimal C:N:P ratio. For example, if the ratio is C:N:P = 120:10:1, then

Moles of nitrogen needed = (moles of carbon present) × (10/120)

Moles of phosphorus needed = (moles of carbon present) × (1/120)

Step 5: Determine the amount of nutrient needed.

Information needed for this calculation:

- Mass of the organic COCs
- Chemical formula of the COCs
- Optimal C:N:P ratio
- Chemical formula of the nutrients

Example 5.23: Determine the Nutrient Requirement for Soil Bioremediation

The results of a feasibility study indicate that the excavated soil in a stockpile (Example 5.22) is suitable for on-site aboveground bioremediation. The feasibility study also determined the optimum C:N:P molar ratio to be 100:10:1. Estimate the amount and cost of nutrients (in lbs) needed to remediate the impacted soil.

Use the following assumptions in your calculation:

- Volume of excavated soil in piles = 375 yd^3
- Initial mass of gasoline in the piles = 158 kg

- Soil porosity = 0.35
- Formula of gasoline (assumed) = C_7H_{16}
- The amounts of N and P naturally occurring in the excavated soil are insignificant
- Trisodium phosphate ($Na_3PO_4 \cdot 12H_2O$) as the P source; price = $10/lb
- Ammonium sulfate ((NH_4)$_2SO_4$) as the N source; price = $3/lb
- One-time nutrient addition only.

Solution:

(a) Determine the number of moles of gasoline:

$$\text{MW of gasoline} = 7 \times 12 + 1 \times 16 = 100$$
$$\text{Moles of gasoline} = 158/100 = 1.58 \text{ kg-mole}$$

(b) Determine the number of moles of C in soil:

Since there are seven carbon atoms in each gasoline molecule, as indicated by its formula, C_7H_{16}, then,

$$\text{Moles of C} = (1.58)(7) = 11.06 \text{ kg-mole}$$

(c) Determine the number of moles of N needed (using the C:N:P ratio):

$$\text{Moles of N needed} = (10/100)(11.06) = 1.106 \text{ kg-mole}$$
$$\text{Moles of } (NH_4)_2SO_4 \text{ needed} = (1.106)/2$$
$$= 0.553 \text{ kg-mole (each mole of ammonium sulfate contains two moles of N)}$$
$$\text{Amount of } (NH_4)_2SO_4 \text{ needed} = (0.553)[(14+4)(2) + 32 + (16)(4)]$$
$$= 73 \text{ kg} = 161 \text{ lb} = \$483$$

(d) Determine the number of moles of P needed (using the C:N:P ratio):

$$\text{Moles of P needed} = (1/100)(11.06) = 0.111 \text{ kg-mole}$$
$$\text{Moles of } Na_3PO_4 \cdot 12H_2O \text{ needed} = 0.111 \text{kg-mole}$$

$$\text{Amount of } Na_3PO_4 \cdot 12H_2O \text{ needed}$$
$$= (0.111)[(23)(3) + 31 + (16)(4) + (12)(18)]$$
$$= 42 \text{ kg} = 92.5 \text{ lb} = \$925$$

Discussion:

The cost of nutrients is relatively low compared to other project expenses.

5.4.4 Oxygen Requirement

For soil bioremediation, the oxygen involved in the biological activity is often supplied through the oxygen in the air. Oxygen is approximately 21% by volume in the ambient air. On the other hand, oxygen is sparingly soluble in water. At 20°C, the saturated dissolved oxygen (DO_{sat}) in water is only about 9 mg/L.

Let us use the following simplified scheme to demonstrate the oxygen requirements:

	C	+	O_2	$\rightarrow$	CO_2
moles:	1		1		1
mass (gram, kg, or lb):	12		32		44

This simplified scheme illustrates that each mole of carbon element requires one mole of oxygen molecule, or every 12 g of carbon requires 32 g of oxygen, a ratio of 2.67. Other elements in the COCs, such as hydrogen, nitrogen, and sulfur, would also demand oxygen for bioremediation. For example, the theoretical amount of oxygen required to aerobically biodegrade benzene can be found as:

	C_6H_6	+	$7.5O_2$	$\rightarrow$	$6CO_2$	+	$3H_2O$
moles:	1		7.5		6		3
mass (gram, kg, or lb):	78		240		264		54

This indicates that each mole of benzene requires 7.5 moles of oxygen molecule, or every 78 g of carbon requires 240 g of oxygen, a ratio of 3.08, which is larger than 2.67 based on pure carbon. Using benzene as the basis, this means that every gram of hydrocarbon requires approximately 3 grams of oxygen for aerobic degradation. It should be noted that this is the theoretical ratio based on the stoichiometric relationship. A larger amount of oxygen would be needed. Using this ratio, the amount of oxygen in an aqueous solution can only support biodegradation of COCs at a concentration of 3 mg/L or less, even if the water is saturated with dissolved oxygen. However, the dissolved oxygen (DO) concentration in the soil moisture would be lower than its saturation value.

Example 5.24: Determine the Oxygen Concentration in Air

Determine the mass concentration of oxygen in ambient air at 20°C. Express the answer in the following units: mg/L, g/L, and lb/ft³.

Solution:

The oxygen concentration in the ambient air is approximately 21% by volume, which is equal to 210,000 ppmV. Equations (2.1) or (2.2) can be used to convert it to a mass concentration,

$$1 \text{ ppmV} = \frac{MW}{24.05} \quad [\text{mg/m}^3] \quad \text{at } T = 20°C$$

$$= \frac{32}{24.05} = 1.33 \text{ mg/m}^3 = 0.00133 \text{ mg/L} \qquad (2.1)$$

or

$$1 \text{ ppmV} = \frac{MW}{385} \times 10^{-6} \quad [\text{lb/ft}^3] \quad \text{at } T = 68°F$$

$$= \frac{32}{385} \times 10^{-6} = 0.083 \times 10^{-6} \text{ lb/ft}^3 \qquad (2.2)$$

Therefore,

$$210,000 \text{ ppmV} = (210,000)(0.00133 \text{ mg/L}) = 279 \text{ mg/L} = 0.28 \text{ g/L}$$
$$= (210,000)(0.083 \times 10^{-6}) = 0.0175 \text{ lb/ft}^3$$

Discussion:

The oxygen concentration in the ambient air, 279 mg/L, is much higher than the saturated dissolved oxygen (DO) concentration in water, 9 mg/L at 20°C.

Example 5.25: Determine the Necessity of Oxygen Addition for *In Situ* Soil Bioremediation

A subsurface contains 5,000 mg/L of gasoline. The air in the subsurface is relatively stagnant. The total bulk density of the soil is 1.8 g/cm³; the degree of water saturation in the soil is 30%; and the porosity is 40%.

Demonstrate that the oxygen in the soil void is not sufficient to support the complete biodegradation of the intruding gasoline.

Solution:

Basis: 1 m³ of soil

(a) Determine the mass of the TPH present:

$$\text{Mass of the soil matrix} = (1 \text{ m}^3)(1{,}800 \text{ kg/m}^3) = 1{,}800 \text{ kg}$$
$$\text{Mass of the TPH} = (5{,}000 \text{ mg/kg})(1{,}800 \text{ kg})$$
$$= 9{,}000{,}000 \text{ mg} = 9{,}000 \text{ g}$$

(b) Use the 3.08 ratio to determine the oxygen requirements for complete oxidation:

$$\text{Oxygen requirement} = (3.08)(9{,}000) = 27{,}720 \text{ g}$$

(c) Determine the amount of oxygen in the soil moisture (assuming that the moisture is saturated with oxygen and the saturated dissolved oxygen concentration in water at 20°C is approximately 9 mg/L):

$$\text{The volume of the soil moisture} = (V)(\phi)(S_w)$$
$$= (1 \text{ m}^3)(40\%)(30\%)$$
$$= 0.12 \text{ m}^3 = 120 \text{ L}$$
$$\text{The amount of oxygen in soil moisture} = (V_1)(DO)$$
$$= (120 \text{ L})(9 \text{ mg/L})$$
$$= 1{,}080 \text{ mg} = 1.08 \text{ g}$$

(d) Determine the amount of oxygen in air (assuming that the oxygen concentration in the pore void is the same as that in the ambient air, 21% by volume, or 279 mg/L from Example 5.24):

$$\text{The volume of the air void, } V_{\text{air void}} = (V)(\phi)(1 - S_w)$$
$$= (1 \text{ m}^3)(40\%)(1 - 30\%)$$
$$= 0.28 \text{ m}^3 = 280 \text{ L}$$
$$\text{The amount of oxygen in air void} = (V_{\text{air void}})(G_{\text{oxygen}})$$
$$= (280 \text{ L})(279 \text{ mg/L})$$
$$= 78{,}120 \text{ mg} = 78.1 \text{ g}$$

(e) The total available oxygen in the soil moisture and the air void
$$= 1.08 + 78.1 = 79.2 \text{ g/m}^3 \text{ soil} \ll 27{,}720 \text{ g/m}^3 \text{ soil}$$

Discussion:

1. The amount of available oxygen in the soil moisture, 1.08 g/m^3 soil, is much smaller than that in the air void, 78.1 g/m^3.

2. It would need at least 255 (i.e., $27{,}720/78.1$) void volumes of fresh air to supply sufficient oxygen for complete biodegradation. The minimum fresh air requirement $= (255)(V_{air\ void}) = (255)(280 \text{ L/m}^3 \text{ soil}) = 71{,}400 \text{ L/m}^3 \text{ soil} = 71.4 \text{ m}^3$ fresh air/m^3 soil.

5.5 Bioventing

5.5.1 Description of the Bioventing Process

Bioventing is an *in situ* soil-remediation technique that uses indigenous microorganisms to biodegrade organic COCs in a subsurface. In bioventing, fresh air is induced into the impacted zone using extraction or injection wells. Oxygen in the air will promote aerobic biodegradation of organic COCs. All aerobically biodegradable COCs can be treated by bioventing. Bioventing is most often used at sites with petroleum products heavier than gasoline (e.g., diesel and jet fuel), while gasoline tends to volatilize readily and can be removed more quickly using soil vapor extraction [7].

5.5.2 Design of the Bioventing Process

A bioventing system is very similar to a soil vapor extraction (SVE) system. It may include vapor extraction well(s), vacuum blower(s), a moisture removal device (the knockout drum), off-gas collection piping and ancillary equipment, and the off-gas treatment system. The main difference between bioventing and SVE is that bioventing promotes biodegradation of COCs and minimizes volatilization. Generally, bioventing uses lower air-flow rates than SVE. The main objectives of the induced air flow are to provide oxygen to promote the biological activities and to carry away metabolic products. The vapor extraction can also be applied intermittently, instead of on a continuous basis, just to provide oxygen to the subsurface. If necessary, nutrient should also be added to the subsurface.

The extent of biological activities can be assessed by determining the carbon dioxide concentration in the extracted air. Excluding the background concentration, carbon dioxide should come from biodegradation of the organic COCs.

Example 5.26: Determine the Efficiency of Bioventing

Bioventing is used to remediate a site impacted by diesel fuel. The average concentrations of TPH and CO_2 in the recently extracted air samples are

500 ppmV and 5%, respectively. Estimate the percentages of diesel removal by volatilization and by biodegradation from this bioventing process.

Strategy:

Same as gasoline, diesel is a mixture of hydrocarbons. Diesel is heavier than gasoline, and the boiling points of diesel compounds range from 200°C to 338°C, compared to 40°C to 205°C for gasoline [7]. Diesel is mainly composed of C_{10}–C_{15} hydrocarbons. In this example, dodecane ($C_{12}H_{26}$) is used to represent diesel, and it has a molecular weight of 170, which is heavier than that of gasoline (100) used earlier in this book.

Carbon dioxide is the dominant greenhouse gas of concern. Even the ambient CO_2 concentration has been increasing; it is still slightly less than 400 ppmV. This background concentration is much smaller than 5% in the extracted air, so it is excluded in the following calculation.

Solution:

(a) MW of diesel $[C_{12}H_{26}] = (12)(12) + (1)(26) = 170$ g/mole
At $T = 20°C$ and $P = 1$ atm,

$$1 \text{ ppmV of diesel} = (\text{MW of diesel}/24.05) \text{ mg/m}^3$$
$$= (170/24.05) \text{ mg/m}^3 = 7.069 \text{ mg/m}^3$$
$$500 \text{ ppmV of diesel} = (500 \text{ ppmV})[7.069 \text{ (mg/m}^3)/\text{ppmV}]$$
$$= (500)(7.069) = 3,534 \text{ mg/m}^3$$

(b) MW of $CO_2 = (12)(1) + (16)(2) = 44$ g/mole
At $T = 20°C$ and $P = 1$ atm,

$$1 \text{ ppmV of } CO_2 = (\text{MW of } CO_2/24.05) \text{ mg/m}^3$$
$$= (44/24.05) \text{ mg/m}^3 = 1.830 \text{ mg/m}^3$$
$$5\% = 50,000 \text{ ppmV of } CO_2$$
$$= (50,000 \text{ ppmV})[1.830 \text{ (mg/m}^3)/\text{ppmV}]$$
$$= (500)(1.830) = 91,500 \text{ mg/m}^3$$

(c) The stoichiometric amount of CO_2 produced from biodegradation of diesel can be found from:

$$C_{12}H_{26} + 18.5O_2 \rightarrow 12CO_2 + 13H_2O$$

Every mole of $C_{12}H_{26}$ biodegraded will produce 12 moles of CO_2. In other words, every gram of CO_2 comes from biodegradation of 0.322 g of diesel [= 170 ÷ (12)(44)].

Therefore, 91,500 mg CO_2/m^3 is equivalent to:

$$(91,500)(0.322) = 29,460 \text{ mg diesel}/m^3$$

(d) Percentage removed by biodegradation
$$= (29,460) \div (29,460 + 3,534) = 89.3\%$$

Percentage removed by volatilization $= (1 - 89.3\%) = 10.7\%$

Discussion:

1. The extracted vapor concentration data suggest that biodegradation accounts for $\approx 90\%$ of the diesel removal.

2. The air-emission rate of diesel will determine if an off-gas treatment system is needed.

Example 5.27: Determine the Rate of Biodegradation from Bioventing

For the bioventing project mentioned in Example 5.26, the air extraction is operated on an intermittent basis. The extraction blower is only on for a consecutive 24-h period every 7 days. As mentioned, the average concentrations of TPH and CO_2 in the recently extracted air samples are 500 ppmV and 5%, respectively. The air extraction rate is equal to 1.0 m^3/min. Estimate the rate of biodegradation occurring in the subsurface and the emission rate of TPH into the atmosphere.

Solution:

(a) As calculated in Example 5.26, 5% CO_2 in the extracted air is equivalent to 91,500 mg CO_2/m^3

$$\text{Rate of } CO_2 \text{ emission} = (Q)(G)$$
$$= (1.0 \text{ } m^3/\text{min})(91,500 \text{ mg } CO_2/m^3)$$
$$= 9,150 \text{ mg/min} = 91.50 \text{ g/min}$$

(b) Total mass of CO_2 emitted during the entire 1,440-min period
$$= (91.5)(1440)$$
$$= 131,760 \text{ g} = 132 \text{ kg } CO_2$$

(c) Total mass of TPH biodegraded over the 7-day period
$$= (132 \text{ kg } CO_2) \times (0.322 \text{ kg TPH/kg } CO_2)$$
$$= 42.5 \text{ kg TPH}$$

(d) The rate of biodegradation over the 7-day cycle
$$= 42.5 \text{ kg} \div 7 \text{ days} = 6.1 \text{ kg/day} = 13.4 \text{ lb/day}$$

(e) The rate of TPH emission = (Q)(G)

$$= (1.0 \text{ m}^3/\text{min})(3{,}534 \text{ mg/m}^3)$$
$$= 3{,}534 \text{ mg/min} = 3.534 \text{ g/min}$$
$$= 5{,}090 \text{ g/day} = 5.09 \text{ kg/day}$$
$$= 11.2 \text{ lb/day}$$

Discussion:

1. To calculate the rate of biodegradation, we should consider the entire 7-day cycle.
2. To determine the air-emission rate of TPH, we may need to consider the instantaneous emission rate.

5.6 *In Situ* Chemical Oxidation

5.6.1 Description of the *In Situ* Chemical Oxidation Process

In situ chemical oxidation (ISCO) involves the introduction of a chemical oxidant into a subsurface to transform COCs in soil or groundwater into less harmful compounds. ISCO is predominantly used to address COCs in the source area so that the mass flux to the groundwater plume can be reduced. Consequently, it can shorten anticipated cleanup times for natural attenuation and other remedial options [7].

5.6.2 Commonly Used Oxidants

There are various oxidants that have been used for ISCO; however, the most commonly used oxidants include:

- Permanganate (MnO_4^-)
- Hydrogen peroxide (H_2O_2)
- Fenton's reagent (hydrogen peroxide + ferrous iron)
- Ozone (O_3)
- Persulfate ($S_2O_8^{2-}$)

The persistence of the oxidant in the subsurface is critical because it affects the extent to which the oxidant can be delivered to the target zone in subsurface. Permanganate can persist for months, persulfate for hours to weeks, and hydrogen peroxide, ozone, and Fenton's reagent can only persist for minutes to hours. Free radicals formed from H_2O_2, $S_2O_8^{2-}$, and O_3 are generally considered to be responsible for transformation of COCs. These intermediates react very quickly and persist for very short periods of time (less than 1 s). Permanganate-based ISCO is more fully developed than the other forms of oxidant [8].

5.6.3 Oxidant Demand

In a chemical oxidation process, the COCs will be oxidized and the oxidant will be reduced. The reaction involves electron transfers in which the oxidant will serve as a terminal electron acceptor by accepting the electrons from the COCs. The half-reactions of some common oxidants are:

$$MnO_4^- + 4H^+ + 3e^- \rightarrow MnO_2 + 2H_2O \tag{5.34}$$

$$H_2O_2 + 2H^+ + 2e^- \rightarrow 2H_2O \tag{5.35}$$

$$2 \cdot OH + 2H^+ + 2e^- \rightarrow 2H_2O \tag{5.36}$$

$$O_3 + 2H^+ + 2e^- \rightarrow O_2 + H_2O \tag{5.37}$$

$$S_2O_8^{2-} + 2e^- \rightarrow 2SO_4^{2-} \tag{5.38}$$

$$\cdot SO_4^- + e^- \rightarrow SO_4^{2-} \tag{5.39}$$

$$O_2 + 4e^- \rightarrow 2O^{2-} \tag{5.40}$$

These equations show that each mole of hydroxyl radical ($\cdot OH$) or sulfate radical ($\cdot SO_4^-$) can accept one mole of electrons; each mole of hydrogen peroxide, ozone, or persulfate can accept two moles of electrons; each mole of permanganate can accept three moles of electrons; and each mole of oxygen can accept four moles of electrons. Table 5.3 tabulates the amount of oxidant needed to transfer one mole of electrons. For a given mass of COC, a smaller oxidant amount would be needed for oxidants that transfer more electrons per unit mass (e.g., oxygen). However, this is not an indicator of whether the

TABLE 5.3

Amount of Oxidant Needed to Transfer One Mole of Electrons

	Electrons Accepted	Molecular Weight	Moles of Electrons Accepted per Unit Mass of Oxidant
Potassium permanganate	3	158	0.0190
Hydrogen peroxide	2	34	0.0588
Ozone	2	48	0.0417
Sodium persulfate	2	238	0.0084
Oxygen	4	32	0.1250

reaction can occur. All the oxidants in this table have a stronger oxidation power than that of oxygen.

To come up with a reaction equation for oxidation of a COC, we need to have another half-reaction to describe the oxidation of the COC. Let us use perchloroethylene (PCE) (C_2Cl_4) as an example:

$$C_2Cl_4 + 4H_2O \rightarrow 2CO_2 + 4Cl^- + 8H^+ + 4e^- \tag{5.41}$$

We then multiply Equation (5.34) by 4 and Equation (5.41) by 3, and add them up to get

$$3C_2Cl_4 + 4MnO_4^- + 4H_2O \rightarrow 6CO_2 + 12Cl^- + 4MnO_2 + 8H^+ \tag{5.42}$$

Equation (5.42) shows that the stoichiometric requirement to oxidize PCE is 4/3 mole permanganate per mole of PCE. Using the same approach, the oxidation of trichloroethylene (TCE) (C_2HCl_3), dichloroethene (DCE) ($C_2H_2Cl_2$), and vinyl chloride (C_2H_3Cl) can be derived as [8]:

$$CHCl_3 + 2MnO_4^- \rightarrow 2CO_2 + 3Cl^- + 2MnO_2 + H^+ \tag{5.43}$$

$$3C_2H_2Cl_2 + 8MnO_4^- \rightarrow 6CO_2 + 6Cl^- + 8MnO_2 + 2OH^- + 2H_2O \tag{5.44}$$

$$3C_2H_3Cl + 10MnO_4^- \rightarrow 6CO_2 + 3Cl^- + 10MnO_2 + 7OH^- + H_2O \tag{5.45}$$

As shown, the stoichiometric requirements for TCE, DCE, and vinyl chloride are 2, 8/3, and 10/3 moles permanganate per mole of COC, respectively. If other oxidants are used, the stoichiometric requirements would be inversely proportional to the ratio of the "electrons accepted" of two oxidants listed in Table 5.3. For example, the stoichiometric requirement of sodium persulfate will be 1.5 times that of potassium permanganate because 1 mole of permanganate can accept 3 moles of electrons, while 1 mole of persulfate can only accept 2 moles of electrons.

In addition to the oxidant demand from COCs, the added oxidants will also be lost due to subsurface reactions unrelated to oxidation of COCs, often referred to as the *natural oxidant demand* (NOD). NOD stems from reactions with organic and inorganic chemical species that are naturally present in the subsurface. Consequently, the total oxidant demand should be the sum of the NOD and the demand from target COCs as:

$$\begin{aligned} \text{Total oxidant demand} = \text{natural oxidant demand} \\ + \text{demand from target COCs} \end{aligned} \tag{5.46}$$

NOD almost always exceeds the oxidant demand from target COCs. NOD has a significant impact in determining if the ISCO is economically feasible and in engineering the applied oxidant dose. Bench- and/or pilot-scale testing should be conducted to determine the NOD for a project.

Example 5.28: Determine the Stoichiometric Amount of Oxidant

The soil at a site is impacted by perchloroethylene (PCE). The soil on top of the capillary fringe contains 5,000 mg/kg of PCE. *In situ* chemical oxidation is considered as one of the remedial alternatives. Determine the stoichiometric amount of potassium permanganate that needs to be delivered to the impacted zone. What would be the amount if sodium persulfate is used?

Solution:

(a) MW of PCE $(C_2Cl_4) = (12)(2) + (35.5)(4) = 166$

   MW of potassium permanganate $(KMnO_4) = (39)(1) + (55)(1) + (16)(4) = 158$

   Concentration of PCE = 5,000 mg/kg = 5.0 g/kg soil

   $= (5.0 \text{ g} \div 166 \text{ g/mole})/\text{kg soil} = 3.01 \times 10^{-2}$ mole PCE/kg soil

   As shown in Equation (5.42), the stoichiometric requirement to oxidize PCE is 4/3 mole permanganate per mole of PCE.

   Stoichiometric amount of $KMnO_4$

   $= (4 \text{ moles of } KMnO_4/3 \text{ moles of PCE}) \times (3.01 \times 10^{-2}$ mole PCE/kg soil)

   $= 4.02 \times 10^{-2}$ mole $KMnO_4$/kg soil

   $= (4.02 \times 10^{-2}$ mole $\times 158$ g/mole $KMnO_4$)/kg soil

   $= 6.35$ g $KMnO_4$/kg soil

(b) MW of sodium persulfate $(Na_2S_2O_8) = (23)(2) + (32)(2) + (16)(8) = 238$

   As shown in Table 5.3 and as discussed previously, the stoichiometric requirement of sodium persulfate will be 1.5 times that of potassium permanganate.

   Stoichiometric amount of $Na_2S_2O_8$

   $= (3 \text{ moles of } Na_2S_2O_8/2 \text{ moles of } KMnO_4) \times (4.02 \times 10^{-2}$ mole $KMnO_4$/kg soil)

   $= (6.02 \times 10^{-2}$ mole $Na_2S_2O_8$/kg soil)

   $= (6.02 \times 10^{-2}$ mole $\times 238$ g/mole $Na_2S_2O_8$/kg soil)

   $= 14.3$ g $Na_2S_2O_8$/kg soil

Example 5.29: Determine the Stoichiometric Amount of Oxidant

The soil at a site is impacted by xylene. The soil on top of the capillary fringe contains 5,000 mg/kg of xylenes. *In situ* chemical oxidation is considered as one of the remedial alternatives. Determine the stoichiometric amount of oxidant that needs to be delivered to the impacted zone.

Solution:

(a) Let us start with oxygen as the oxidant:

$$C_6H_4(CH_3)_2 + 10.5O_2 \rightarrow 8CO_2 + 5H_2O$$

As shown in this equation, the stoichiometric requirement is 10.5 moles of oxygen per mole of xylenes. To express the oxygen requirement on the basis of g O_2/g xylenes:

$$MW \text{ of } C_6H_4(CH_3)_2 = (12)(8) + (1)(10) = 106$$

Concentration of xylenes = 5,000 mg/kg = 5.0 g/kg soil

= (5.0 g ÷ 106 g/mole)/kg soil = 4.72×10^{-2} mole xylenes/kg soil

Stoichiometric amount of O_2 (using the molar ratio)

= (10.5 moles of O_2/mole of xylenes) × (4.72×10^{-2} mole xylenes/kg soil)

= 0.495 mole O_2/kg soil

= (0.495 mole × 32 g/mole O_2)/kg soil

= 15.85 g O_2/kg soil

Oxygen requirement (in mass ratio)

= (10.5 moles of O_2/mole of xylenes) × [(32 g/mole) ÷ (106 g/mole)]

= 3.17 g O_2/g xylenes

Stoichiometric amount of O_2 (using the mass ratio)

= (3.17 g O_2/g xylenes) × (5.0 g xylenes/kg soil)

= 15.85 g O_2/kg soil

(b) Now let us determine the stoichiometric amount of sodium persulfate, if it is the oxidant to be applied.

$$MW \text{ of sodium persulfate } (Na_2S_2O_8) = (23)(2) + (32)(2) + (16)(8) = 238$$

As shown in Table 5.3 and as previously discussed, the stoichiometric requirement of sodium persulfate will be two times that of oxygen.

Stoichiometric amount of $Na_2S_2O_8$ (using the molar ratio)

= (2 moles of $Na_2S_2O_8$/mole of O_2) × (0.495 mole O_2/kg soil)

= (0.99 mole $Na_2S_2O_8$/kg soil)

= (0.99 mole × 238 g/mole $Na_2S_2O_8$/kg soil)

= 236 g $Na_2S_2O_8$/kg soil

The stoichiometric requirements would be inversely proportional to the ratio of the "moles of electrons accepted per unit
mass of oxidant" of two oxidants listed in Table 5.3.

Stoichiometric amount of $Na_2S_2O_8$ (using the mass ratio)

= (15.85 g O_2/kg soil) × (0.125/0.0084)

= 236 g $Na_2S_2O_8$/kg soil

Discussion:

1. The stoichiometric oxygen requirements for typical petroleum
   hydrocarbons range from 3.0 to 3.5 g O_2/g COC.
2. The stoichiometric amounts of other oxidants can be readily
   found from that of oxygen by using the molar ratio or the mass
   ratio.

5.7 Thermal Destruction

5.7.1 Description of the Thermal Destruction Process

Thermal destruction, considered here, is an *ex situ* remediation technique
to remediate soil impacted by organics. *Ex situ* thermal treatment generally involves destruction or removal of organic COCs through exposure
to high temperature in treatment cells, combustion chambers, or other
means used to contain the impacted media during the treatment. There are
many different thermal treatment alternatives available, including thermal
destruction/oxidation, pyrolysis, vitrification, thermal desorption, plasma
high-temperature recovery, infrared, and wet-air oxidation. This section
focuses on thermal destruction/oxidation (or combustion).

The common combustion units used for hazardous wastes are incinerators, boilers, and industrial furnaces. During combustion, organic wastes
are converted into gases. The stable gases produced from combustion of
organics are primarily carbon dioxide and water vapor. However, small
quantities of carbon monoxide, hydrogen chloride, and other gases may
form. These gases have potential adverse impacts to human health and the
environment [9].

5.7.2 Design of the Combustion Units

The key design components of combustion units are the three Ts, which are
combustion temperature, residence time (also called "dwell time"), and *turbulence.* They affect the size of a reactor and its destruction efficiency. Other
important parameters to be considered include heating value of the influent
and the requirements of auxiliary fuel and supplementary air.

Organic compounds generally contain heating values. These organic compounds can also serve as energy sources for combustion. The higher the organic concentration in a waste stream, the higher the heat content is and the lower the requirement of auxiliary fuel would be. If the energy content in a waste stream is greater than 4,000 Btu/lb, it can sustain burning without supplementary fuel. If the heating value of a compound is not available, the following Dulong's formula can be used:

$$\text{Heating value (in Btu/lb)} = 145.4C + 620\left(H - \frac{O}{8}\right) + 41S \qquad (5.47)$$

where C, H, O, and S are the percentages by weight of these elements in the compound.

To ensure a more complete combustion, excess air should be provided in addition to the stoichiometric amount for combustion of the COCs. Combustion temperature should be high enough to achieve the required destruction efficiency. The higher the combustion temperature, the shorter the required residence time would be for the specified destruction efficiency. The combustion temperature (T in °F) can be estimated by [10]:

$$T = 60 + \frac{NHV}{(0.325) \times [1 + (1 + EA)(7.5 \times 10^{-4})(NHV)]} \qquad (5.48)$$

where NHV is the net heating value in Btu/lb and EA is the excess air in percent.

Example 5.30: Determine the Energy Content of a Waste Sample

Several leaky USTs, which previously stored xylenes, $C_6H_4(CH_3)_2$, were removed. The excavated soil was stockpiled on site and contains an average xylene concentration of 1,500 mg/kg. The volume of the stockpile is 500 m³. The excavated soil needs to be treated before final disposal. Direct incineration of the soil is considered as a remedial alternative. Assuming the original organic content of the native soil is negligible, show that the heating value of the soil containing 1,500 mg/kg xylenes is very low.

Solution:

(a) The carbon content in the soil containing 1,500 mg/kg xylenes (C_8H_{10}) =

$$1,500 \times \frac{\text{Mass of C in one mole of xylenes}}{\text{MW of xylenes}}$$

$$= 1,500 \times \frac{12 \times 8}{(12 \times 8 + 1 \times 10)} = 0.136\%$$

(b) The hydrogen content in the soil containing 1,500 mg/kg xylenes (C_8H_{10}) =

$$1,500 \times \frac{\text{Mass of H in one mole of xylenes}}{\text{MW of xylenes}}$$

$$= 1,500 \times \frac{1 \times 10}{(12 \times 8 + 1 \times 10)} = 0.014\%$$

(c) The heating value (using Equation 5.47):

$$\text{Heating value}\left(\text{in}\,\frac{\text{Btu}}{\text{lb}}\right) = 145.4\,(0.136) + 620(0.014) = 28.5$$

Example 5.31: Estimate the Combustion Temperature

A waste stream contains 20% by weight of carbon, 2% of oxygen, 1% of hydrogen, and 0.1% of sulfur. Estimate the combustion temperature with no auxiliary fuel and 85% excess air.

Solution:

(a) The heating value (using Equation 5.47):

$$\text{Heating value}\left(\text{in}\,\frac{\text{Btu}}{\text{lb}}\right) = 145.4\,(20) + 620\left(1 - \frac{2}{8}\right) + 41(0.1)$$

$$= 3,377$$

(b) The combustion temperature (using Equation 5.48) =

$$T = 60 + \frac{3,377}{(0.325) \times [1 + (1 + 85\%)(7.5 \times 10^{-4})(3,377)]}$$

$$= 1,828°F$$

5.7.3 Regulatory Requirements for Incineration of Hazardous Waste

Emissions from hazardous waste combustors are regulated under two statutory authorities: Resource Conservation and Recovery Act (RCRA) and Clean Air Act (CAA). The maximum achievable control technology (MACT)

standards set emission limitations for dioxins/furans, heavy metals, par-
ticulates, hydrogen chloride, hydrocarbons and carbon monoxide (CO), and
destruction and removal efficiency (DRE) for organics. Because the primary
purpose of a combustion unit is to destroy the organics in the hazardous
waste, the combustion unit must demonstrate a DRE of 99.99% for each prin-
cipal organic hazardous constituent (POHC) in the hazardous waste stream.
The required DRE for certain dioxin-containing wastes is even greater, at
99.9999% [9]. The DRE (in %) is defined as

$$\text{DRE} = \frac{M_{in} - M_{out}}{M_{in}} \tag{5.49}$$

where
M_{in} = feed rate of a particular POHC to the combustion unit (lb/h or
   kg/h)
M_{out} = output rate of a particular POHC from the combustion unit (lb/h or
   kg/h)

The exhaust from the combustion unit is usually continuously monitored
and recorded for various constituents, including CO. The combustion efficiency is calculated as follows, and it is typically required to be >99.90%:

$$\text{Combustion efficiency} = \frac{[CO_2]}{[CO_2] + [CO]} \times 100\% \tag{5.50}$$

where
$[CO_2]$ = CO_2 concentration in the exhaust on a dry basis (in ppmV)
$[CO]$ = CO concentration in the exhaust on a dry basis (in ppmV)

Example 5.32: Determine the Destruction and Removal Efficiencies

The results of a trial burn (T = 2,000°F and residence time = 30 s) of a combustion unit on a waste stream consisting of three POHCs are shown in the
following grid:

	Feed (kg/h)	Outlet (kg/h)
Benzene (C_6H_6)	500	0.04
Phenol (C_6H_5OH)	300	0.04
PCE (C_2Cl_4)	200	0.01

Is this unit in compliance?

Solution:

(a) DRE of benzene:

$$DRE = \frac{500 - 0.04}{500} = 99.992\% > 99.99\%$$

(b) DRE of phenol:

$$DRE = \frac{300 - 0.04}{300} = 99.987\% < 99.99\%$$

(c) DRE of PCE:

$$DRE = \frac{200 - 0.01}{200} = 99.995\% > 99.99\%$$

The combustion unit is not in compliance because DRE of phenol is less than 99.99%.

Example 5.33: Determine the Generation of Hydrogen Chloride

For the trial burn mentioned in Example 5.32, if all the chlorine in the feed stream is converted to hydrogen chloride, estimate the hydrogen chloride flow rate prior to emission control.

Solution:

(a) MW of PCE (C_2Cl_4) = (12)(2) + (35.5)(4) = 166
Molar flow rate of PCE = (200,000 g/h) ÷ (166 g/mole) = 1,205 moles/h

(b) Molar flow rate of HCl = molar flow rate of Cl = 2 × molar flow rate of PCE
= 2 × 1,205 = 2,410 moles/h

(c) MW of HCl = 1 + 35.5 = 36.5
Mass flow rate of HCl = (2,410 moles/h) × (36.5 g/mole) = 97,965 g/h
= 97.97 kg/h

Discussion:

The general RCRA requirement for HCl emission is ≤1.8 kg/h or >99% reduction. In this case, a removal of HCl is required.

Example 5.34: Determine the Combustion Efficiency

The off-gas from a combustion unit was analyzed by an Orsat gas analyzer. The off-gas (on a dry basis) consisted of 17% CO_2, 2.5% O_2, 80% N_2, and 160 ppmV of CO. Estimate the combustion efficiency of this combustion unit.

Solution:

As mentioned in Chapter 2, 1% = 10,000 ppmV

From Equation (5.50),

$$\text{Combustion efficiency} = \frac{17,000}{17,000+160} \times 100\% = 99.06\%$$

Discussion:

The calculated combustion efficiency is <99.9%. A better mixing, more excess air, or higher combustion temperature may be needed to raise the combustion efficiency above 99.9%.

5.8 Low-Temperature Thermal Desorption

5.8.1 Description of the Low-Temperature Thermal Desorption Process

Low-temperature thermal desorption (LTTD), also known as low-temperature thermal heating, low-temperature thermal volatilization, and thermal stripping, is an *ex situ* soil remediation technique. In the low-temperature thermal desorption process, volatile and semi-volatile COCs are removed from soil, sediments, or slurries through volatilization that is enhanced by elevated temperatures. The process is typically operated at temperatures from 200°F up to 1,000°F. The term *low temperature* is used to differentiate the process from incineration. At these lower temperatures, the COCs are physically driven off from the soil matrix instead of being combusted. The produced off-gas requires further treatment before being vented to atmosphere.

5.8.2 Design of the Low-Temperature Thermal Desorption Process

There are no set guidelines for design of a low-temperature heating reactor. The time required to achieve a specific final concentration would depend mainly on the following factors:

- *Temperature inside the reactor*: The higher the temperature, the higher the desorption rate will be and, consequently, the shorter the retention time.

- *Mixing conditions inside the reactor*: Better mixing conditions will enhance the heat transfer and improve venting of the desorbed COCs.
- *Volatility of the COCs*: The more volatile the COCs are, the shorter the required retention time will be.
- *Size of the soil particles*: The smaller the soil particles, the easier the desorption will be.
- *Types of soil*: Clay has a stronger affinity with COCs and, thus, the COCs will be harder to desorb from clayey material.

The rate of desorption or the required detention time to remediate a specific type of soil to a permissible concentration can be best determined from a pilot study. The results from the pilot study should then be used for the preliminary design of the full-scale operation. The desorption process can be conducted in a batch mode or in a continuous mode. For the continuous mode, the reactor can be modeled as a continuous-flow stirred tank reactor (CFSTR) if the soil is relatively well-mixed inside the reactor. For the desorption reaction, a first-order type of reaction is a reasonable assumption. For a first-order reaction, the relationship among the influent and final concentrations, reaction rate constant, and residence time are as follows (see Chapter 4 for more detailed discussions):

Batch reactor

$$\frac{C_f}{C_i} = e^{-k\tau} \quad \text{or} \quad C_f = (C_i)e^{-k\tau} \tag{4.16}$$

CFSTR

$$\frac{C_{out}}{C_{in}} = \frac{1}{1 + k(V/Q)} = \frac{1}{1 + k\tau} \tag{4.20}$$

Example 5.35: Determine the Residence Time for Low-Temperature Heating (Batch Mode of Operation)

A batch-type low-temperature thermal desorption reactor is proposed to treat soil containing 2,500 mg/kg of total petroleum hydrocarbon (TPH). A pilot study was conducted, and it took 25 min to reduce the concentration to 150 mg/kg. First-order kinetics applies. If the required final soil TPH concentration is 50 mg/kg, what should be the design residence time of the soil in the reactor?

Solution:

(a) Determine the rate constant by using Equation (4.16):

$$\frac{C_f}{C_i} = e^{-kt} = \frac{150}{2,500} = e^{-k(25)}$$

So, $k = 0.113/\text{min}$

(b) Now, we use this rate constant and Equation (4.16) to determine the required detention time:

$$\frac{C_f}{C_i} = e^{-kt} = \frac{50}{2,500} = e^{-(0.113)\tau}$$

So, $\tau = 35$ min

Discussion:

The rate constant is often obtained from bench-scale experiments by using the batch-type reactors.

Example 5.36: Determine the Residence Time for Low-Temperature Thermal Desorption (Continuous Mode of Operation)

A low-temperature thermal-desorption soil reactor is proposed to treat soil containing 2,500 mg/kg of total petroleum hydrocarbon (TPH) in a continuous mode of operation. Assume that the reactor is a CFSTR and that first-order kinetics applies. A pilot study was conducted and the reaction-rate constant was determined to be 0.3/min. The required final soil TPH concentration is 100 mg/kg.

1. What should be the design residence time of the soil in the reactor?
2. The soil content of the reactor is to be kept at less than 30% of the total reactor volume to allow for efficient mixing. Estimate the required size of the reactor vessel to treat the impacted soil at a rate of 500 kg/h.

Solution:

(a) Determine the required retention time by using Equation (4.20):

$$\frac{C_{out}}{C_{in}} = \frac{1}{1 + k\tau} = \frac{100}{3,000} = \frac{1}{1 + (0.3)\tau}$$

$$1 + 0.3\tau = 30$$

So, $\tau = 97$ min $= 1.61$ h

(b)   Assuming the bulk density of soil in the reactor is 1.8 g/cm^3, the volumetric feeding rate of the soil can be found as:

$$Q_{soil} = (500 \text{ kg/h}) \div 1.8 \text{ kg/L} = 278 \text{ L/h}$$

The minimum reactor size can be found from the definition of the retention time as:

$$\tau = V/Q = 1.61 \text{ h} = V/(278 \text{ L/h})$$

So, $V = 447$ L

With the soil occupying less than 30% of the total reactor volume, the required reactor volume ($V_{reactor}$) can be found as

$$V_{reactor} = (447) \div 30\% = 1,490 \text{ L} = 394 \text{ gal}$$

References

1. Johnson, P.C., and R.A. Ettinger. 1994. Considerations for the design of *in situ* vapor extraction systems: Radius of influence vs. zone of remediation. *Groundwater Monitoring and Remediation* 14 (3): 123–28.
2. Johnson, P.C., M.W. Kemblowski, and J.D. Colthart. 1990. Qualitative analysis for the cleanup of hydrocarbon-contaminated soils by *in situ* soil venting. *Groundwater* 28 (3): 413–29.
3. Johnson, P.C., C.C. Stanley, M.W. Kemblowski, D.L. Byers, and J.D. Colthart. 1990. A practical approach to the design, operation, and monitoring of *in situ* soil-venting systems. *Groundwater Monitoring and Remediation* 10 (2): 159–78.
4. Kuo, J.F., E.M. Aieta, and P.H. Yang. 1991. Three-dimensional soil venting model and its applications. In *Emerging technologies in hazardous waste management II*, ed. D.W. Tedder and F.G. Pohland, 382–400. American Chemical Society Symposium Series 468. Washington, DC: ACS.
5. Peters, M.S., and K.D. Timmerhaus. 1991. *Plant design and economics for chemical engineers*. 4th ed. New York: McGraw-Hill.
6. USEPA. 1991. Site characterization for subsurface remediation. EPA/625/R-91/026. Washington, DC: Office of Research and Development, US EPA.

7. USEPA. 2004. How to evaluate alternative cleanup technologies for underground storage sites. EPA/510/R-04/002, Washington, DC: Office of Solid Waste and Emergency Response, US EPA.
8. USEPA. 2006. *In situ* chemical oxidation. EPA/600/R-06/002. Washington, DC: Office of Research and Development, US EPA.
9. USEPA. 2011. RCRA orientation manual 2011: Resource Conservation and Recovery Act. EPA/530/F-11/003. Washington, DC: Office of Solid Waste and Emergency Response, US EPA.
10. Santoleri, J., J. Reynolds, and L. Theodore. 2004. *Introduction to hazardous waste incineration*. 2nd ed. Hoboken, NJ: John Wiley & Sons.

6

Groundwater Remediation

6.1 Introduction

When an aquifer is impacted, groundwater extraction is often needed. Groundwater extraction mainly serves two purposes: (1) to minimize the plume migration or spreading, and (2) to reduce the concentrations of compounds of concern (COCs) in the impacted aquifer. The extracted water often needs to be treated before being put back into the aquifer or released to surface water. *Pump-and-treat* is a general term used in groundwater remediation that extracts impacted groundwater and treats it aboveground.

Groundwater extraction is typically accomplished through one or more pumping or extraction wells. Extraction of groundwater stresses the aquifer and creates a cone of depression or a capture zone. Deciding spacing among the extraction wells and choosing appropriate locations for the wells are important components in design. Extraction wells should be strategically located to accomplish rapid mass removal from areas of the groundwater plume where COCs are heavily concentrated. On the other hand, they should be located to allow full capture of the plume to prevent further migration/spreading. In addition, if containment of the plume is the main objective for the groundwater extraction, the pumping rate could be established at a minimal rate, just sufficient to prevent the plume migration, since the extracted groundwater often incurs treatment costs. On the other hand, if the groundwater cleanup is required, the extraction rate may need to be enhanced to shorten the remediation time. For both cases, major questions to be answered for design of a groundwater remediation program are similar, which include

1. What is the optimum number of extraction wells?
2. Where would be the optimal locations of the extraction wells?
3. What would be the sizes (diameters) of the wells?
4. What would be the well depth and the interval and size of the perforations?

5. What would be the construction materials of the wells?
6. What would be the optimum pumping rate of each well?
7. What would be the optimal treatment method for the extracted groundwater?
8. What would be the disposal options for the treated groundwater?

An impacted aquifer can also be remediated using *in situ* technologies. This chapter starts with design calculations for the capture zone and optimal well spacing. The rest of the chapter focuses on design calculations for commonly used *in situ* and *ex situ* groundwater remediation techniques, including activated-carbon adsorption, air stripping, *ex situ* and *in situ* bioremediation, air sparging, biosparging, chemical precipitation, *in situ* chemical oxidation, and advanced oxidation.

6.2 Groundwater Extraction

This section covers common calculations for determining the zone of influence of a groundwater extraction system. The results from these calculations can provide answers to some of the questions mentioned previously.

6.2.1 Cone of Depression

When a groundwater extraction well is pumped, the water level in its vicinity will decline, thereby providing a gradient to drive water toward the well. The gradient is steeper as the well is approached, and this results in a cone of depression. In dealing with groundwater extraction, evaluation of the cone of depression of a pumping well is critical because it represents the limit that the well can reach.

The equations describing the steady-state flow of an aquifer into a fully penetrating well have been discussed in Section 3.3. The equations were used in that section to estimate both the drawdown in the wells and the hydraulic conductivity of the aquifer. These equations can also be used to estimate the radius of influence of a groundwater extraction well and/ or the groundwater extraction rate. This subsection will illustrate these applications.

6.2.1.1 Steady-State Flow in a Confined Aquifer

Equation (6.1) describes steady-state flow from a fully penetrating well in a confined aquifer (an artesian aquifer). A fully penetrating well means that the groundwater can enter at any level from the top to the bottom of the aquifer.

$$Q = \frac{Kb(h_2 - h_1)}{528 \log(r_2/r_1)} \quad \text{(American Practical Units)}$$

$$= \frac{2.73\, Kb(h_2 - h_1)}{\log(r_2/r_1)} \quad \text{(SI Units)} \tag{6.1}$$

where

Q = pumping rate or well yield (gpm or m^3/day)
h_1, h_2 = static head measured from the aquifer bottom (ft or m)
r_1, r_2 = radial distance from the pumping well (ft or m)
b = thickness of the aquifer (ft or m)
K = hydraulic conductivity of the aquifer (gpd/ft^2 or m/day)

Example 6.1: Steady-State Drawdown from Pumping a Confined Aquifer

A confined aquifer 30-ft (9.1-m) thick has a piezometric surface 80 ft (24.4 m) above the bottom confining layer. Groundwater is being extracted from a 4-in. (0.1-m) diameter fully penetrating well.

The pumping rate is 40 gpm (0.15 m^3/min). The aquifer is relatively sandy and has a hydraulic conductivity of 200 gpd/ft^2. Steady-state drawdown of 5 ft (1.5 m) is observed in a monitoring well 10 ft (3.0 m) from the pumping well. Determine

(a) The drawdown in the pumping well
(b) The radius of influence of the pumping well

Solution:

(a) First let us determine h_1 (at $r_1 = 10$ ft):

$h_1 = 80 - 5 = 75$ ft (or $= 24.4 - 1.5 = 22.9$ m)

To determine the drawdown in the pumping well, set r at the well = well radius = (2/12) ft = 0.051 m and use Equation (6.1):

$$40 = \frac{(200)(30)(h_2 - 75)}{528 \log[(2/12)/10]} \quad \rightarrow h_2 = 68.74 \text{ ft}$$

or

$$[(0.15)(1,440)] = \frac{2.73\,[(200)(0.0410)](9.1)(h_2 - 22.9)}{\log(0.051/3.0)} \quad \rightarrow h_2 = 21.0 \text{ m}$$

So the drawdown in the pumping well = $80 - 68.7 = 11.3$ ft (or $= 24.4 - 21.0 = 3.4$ m)

(b) To determine the radius of influence of the pumping well, set the radius of influence (r_{RI}) at the location where the drawdown is equal to zero and use the drawdown information of the extraction well:

$$40 = \frac{(200)(30)(68.7 - 80)}{528 \log[(2/12)/r_{RI}]} \qquad \rightarrow r_{RI} = 270 \text{ ft}$$

or

$$[(0.15)(1,440)] = \frac{2.73\,[(200)(0.041)](9.1)(21.0 - 24.4)}{\log(0.051/r_{RI})} \qquad \rightarrow r_{RI} = 82 \text{ m}$$

Similar results can also be derived from using the drawdown information of the observation well ($r = 10$ ft = 3 m) as:

$$40 = \frac{(200)(30)(75 - 80)}{528 \log[10/r_{RI}]} \qquad \rightarrow r_{RI} = 263 \text{ ft}$$

or

$$[(0.15)(1,440)] = \frac{2.73\,[(200)(0.0410)](9.1)(22.9 - 24.4)}{\log(3/r_{RI})} \qquad \rightarrow h_2 = 78 \text{ m}$$

Discussion:

1. In parts (a) and (b), 0.041 is the conversion factor to convert the hydraulic conductivity from gpd/ft^2 to m/day. The factor was taken from Table 3.1.

2. Calculations in part (a) have demonstrated that the results would be the same by using two different systems of units.

3. The h_1-h_2 term can be replaced by s_2-s_1, where s_1 and s_2 are the drawdown values at r_1 and r_2, respectively.

4. The differences in the calculated r_{RI} values in part (b) come mainly from the unit conversions and data truncations.

Example 6.2: Estimate the Groundwater Extraction Rate from a Confined Aquifer Using the Steady-State Drawdown Data

Use the following information to estimate the groundwater extraction rate of a fully penetrating well in a confined aquifer:

- Aquifer thickness = 30.0-ft (9.1-m) thick
- Well diameter = 4-in. (0.1-m) diameter
- Well perforation depth = fully penetrating

- Hydraulic conductivity of the aquifer = 400 gpd/ft²
- Steady-state drawdown

= 2.0 ft in a monitoring well that is 5 ft from the pumping well
= 1.2 ft in a monitoring well that is 20 ft from the pumping well

Solution:

Inserting the data into Equation (6.1), we obtain

$$Q = \frac{Kb(h_2 - h_1)}{528 \log(r_2/r_1)} = \frac{(400)(30)(2.0 - 1.2)}{528 \log(20/5)} = 30.2 \text{ gpm}$$

Discussion:

The $h_2 - h_1$ term can be replaced by $s_2 - s_1$, where s_1 and s_2 are the drawdown values at r_1 and r_2, respectively.

6.2.1.2 Steady-State Flow in an Unconfined Aquifer

The equation describing the steady-state flow from a fully penetrating well in an unconfined aquifer (water-table aquifer) can be expressed as:

$$Q = \frac{K(h_2^2 - h_1^2)}{1,055 \log(r_2/r_1)} \quad \text{(American Practical Units)}$$

$$= \frac{1.366\, K(h_2^2 - h_1^2)}{\log(r_2/r_1)} \quad \text{(SI Units)} \tag{6.2}$$

All the terms are as defined for Equation (6.1).

Example 6.3: Steady-State Drawdown from Pumping an Unconfined Aquifer

A water-table aquifer is 40-ft (12.2-m) thick. Groundwater is being extracted from a 4-in. (0.1-m) diameter fully penetrating well.

The extraction rate is 40 gpm (0.15 m³/min). The aquifer is relatively sandy and has a hydraulic conductivity of 200 gpd/ft². Steady-state drawdown of 5 ft (1.5 m) is observed in a monitoring well at 10 feet (3.0 m) from the pumping well. Determine

(a) The drawdown in the pumping well
(b) The radius of influence of the pumping well

Solution:

(a) First let us determine h_1 (at $r_1 = 10$ ft):

$h_1 = 40 - 5 = 35$ ft (or = 12.2 − 1.5 = 10.7 m)

To determine the drawdown in the pumping well, set r at the pumping well = well radius = $(2/12)$ ft = 0.051 m, and use Equation (6.2):

$$40 = \frac{(200)(h_2^2 - 35^2)}{1,055 \log[(2/12)/10]} \quad \rightarrow h_2 = 29.2 \text{ ft}$$

or

$$[(0.15)(1440)] = \frac{1.366\,[(200)(0.0410)](h_2^2 - 10.7^2)}{\log(0.051/3.0)} \quad \rightarrow h_2 = 9.0 \text{ m}$$

So the drawdown in the extraction well = $40 - 29.2 = 10.8$ ft (or $= 12.2 - 9.0 = 3.2$ m)

(b) To determine the radius of influence of the pumping well, set r at the radius of influence (r_{RI}) at the location where the drawdown is equal to zero. We can use the drawdown information of the pumping well as:

$$40 = \frac{(200)(29.2^2 - 40^2)}{1,055 \log[(2/12)/r_{RI}]} \quad \rightarrow r_{RI} = 580 \text{ ft}$$

or

$$[(0.15)(1440)] = \frac{1.366\,[(200)(0.0410)](9.0^2 - 12.2^2)}{\log(0.051/r_{RI})} \quad \rightarrow r_{RI} = 168 \text{ m}$$

Similar results can also be derived from using the drawdown information of the observation well as:

$$40 = \frac{(200)(35^2 - 40^2)}{1,055 \log[10/r_{RI}]} \quad \rightarrow r_{RI} = 598 \text{ ft}$$

or

$$[(0.15)(1440)] = \frac{1.366\,[(200)(0.0410)](10.7^2 - 12.2^2)}{\log(3/r_{RI})} \quad \rightarrow r_{RI} = 181 \text{ m}$$

Discussion:

1. As discussed in Example 6.2, the $h_2 - h_1$ term in Equation (6.1) (for confined aquifers) can be replaced by $s_2 - s_1$. However, no analogy can be made here, that is, $h_2^2 - h_1^2$ in Equation (6.2) cannot be replaced by $s_1^2 - s_2^2$.

2. The differences in the calculated r_{RI} values in part (b) come mainly from the unit conversions and data truncations.

Example 6.4: Estimate the Groundwater Extraction Rate from an Unconfined Aquifer Using the Steady-State Drawdown Data

Use the following information to estimate the groundwater extraction rate of a pumping well in an unconfined aquifer:

- Aquifer thickness = 30.0-ft (9.1-m) thick
- Well diameter = 4-in. (0.1-m) diameter
- Well perforation depth = fully penetrating
- Hydraulic conductivity of the aquifer = 400 gpd/ft²
- Steady-state drawdown

    = 2.0 ft in a monitoring well that is 5 ft from the pumping well

    = 1.2 ft in a monitoring well that is 20 ft from the pumping well

Solution:

(a) First we need to determine h_1 and h_2:

$h_1 = 30.0 - 2.0 = 28.0$ ft

$h_2 = 30.0 - 1.2 = 28.8$ ft

(b) Inserting the data into Equation (6.2), we obtain

$$Q = \frac{K(h_2^2 - h_1^2)}{1,055 \log(r_2/r_1)} = \frac{400(28.8^2 - 28.0^2)}{1,055 \log(20/5)} = 28.6 \text{ gpm}$$

6.2.2 Capture-Zone Analysis

One key element in design of a groundwater extraction system is to select proper locations for the extraction wells. If only one well is used, the well should be strategically located to create a capture zone that encloses the entire plume. If two or more wells are used, the general interest is to find the maximum distance between two consecutive wells such that no COCs can escape through the interval between these two wells. Once such distances are determined, one can depict the capture zone of these wells.

To delineate the capture zone of a groundwater pumping system in an actual aquifer can be very complicated. To allow for a theoretical approach, let us consider a homogeneous and isotropic aquifer with a uniform thickness and assume that the groundwater flow is uniform and steady. The theoretical treatment of this subject starts from one single well and expands to multiple wells. The following discussion is mainly based on the work by Javandel and Tsang [1].

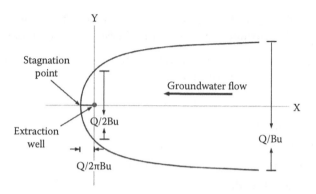

FIGURE 6.1
Capture zone of one extraction well.

6.2.2.1 One Groundwater Extraction Well

For easier presentation, let the extraction well be located at the origin of an *x-y* coordinate system (Figure 6.1). The equation of the dividing streamlines that separate the capture zone of this well from the rest of the aquifer (sometimes referred as the *envelope*) is

$$y = \pm \frac{Q}{2Bu} - \frac{Q}{2\pi\, Bu} \tan^{-1} \frac{y}{x} \qquad (6.3)$$

where
  B = aquifer thickness (ft or m)
  Q = groundwater extraction rate (ft^3/s or m^3/s)
  u = groundwater velocity (ft/s or m/s) = Ki
  Figure 6.1 illustrates the capture zone of one pumping well. The larger the Q/Bu value (i.e., larger groundwater extraction rate, slower groundwater velocity, or shallower aquifer thickness), the larger the capture zone will be. Three interesting sets of x and y values of the capture zone:

1. The stagnation point, where y is approaching zero
2. The sidestream distances at the line of the extraction well, where $x = 0$
3. The asymptotic values of y, where $x = \infty$

If these three sets of data are determined, the rough shape of the capture zone can be depicted. At the stagnation point (where y is approaching zero), the distance between the stagnation point and the pumping well is equal to $Q/2\pi Bu$, which represents the farthest downstream distance that the pumping well can reach. At $x = 0$, the maximum sidestream distance from the extraction well is equal to ±$Q/4Bu$. In other words, the distance between the dividing streamlines at the line of the well is equal to $Q/2Bu$. The asymptotic value of y (where $x = \infty$) is equal to ±$Q/2Bu$. Thus, the distance between the streamlines far upstream from the pumping well is Q/Bu.

Note that the parameter in Equation (6.3), (Q/Bu), has a dimension of length. To draw the envelope of the capture zone, Equation (6.3) can be rearranged as:

$$x = \frac{y}{\tan\left\langle[+1-(\frac{2Bu}{Q})y]\pi\right\rangle} \quad \text{for positive } y \text{ values} \tag{6.4}$$

$$x = \frac{y}{\tan\left\langle[-1-(\frac{2Bu}{Q})y]\pi\right\rangle} \quad \text{for negative } y \text{ values} \tag{6.5}$$

A set of (x, y) values can be obtained from these equations by first specifying a value of y. The envelope is symmetrical about the x-axis.

Example 6.5: Draw the Envelope of a Capture Zone of a Groundwater Pumping Well

Delineate the capture zone of a groundwater recovery well with the following information:

- $Q = 60$ gpm
- Hydraulic conductivity = 2,000 gpd/ft^2
- Groundwater gradient = 0.01
- Aquifer thickness = 50 ft

Solution:

(a) Determine the groundwater velocity, u:

$u = (K)(i) = [(2,000 \text{ gpd/ft}^2)(1 \text{ day}/1,440 \text{ min})(1 \text{ ft}^3/7.48 \text{ gal})](0.01)$
$= 1.86 \times 10^{-3}$ ft/min

(b) Determine the value of the parameter, Q/Bu:

$$\frac{Q}{Bu} = \frac{(60 \text{ gal/min})(1 \text{ ft}^3/7.48 \text{ gal})}{(50 \text{ ft})(1.86 \times 10^{-3} \text{ ft/min})}$$

$$= \frac{60 \text{ gal/min}}{(50 \text{ ft})[(2,000 \text{ gal/d/ft}^2)(1 \text{ d}/1,440 \text{ min})(0.01)]} = 86.4 \text{ ft}$$

(c) Establish a set of (x, y) values using Equation (6.4). First specify values of y (select smaller intervals for small y values).

The following grid lists some of the data points used to plot Figure 6.2.

y (ft)	x (ft)
0	0.00
0.1	−13.74
1	−13.73
5	−13.14
10	−11.24
20	−2.34
30	21.01
40	168.78
−0.1	−13.74
−1	−13.73
−5	−13.14
−10	−11.24
−20	−2.34
−30	21.01
−40	168.78

See Figure 6.2.

Discussion:

1. The capture-zone curve is symmetrical about the *x*-axis, as shown in the table or in the figure. Note that Equation (6.4) should be used for positive *y* values and Equation (6.5) for negative *y* values.

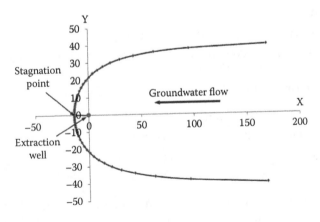

FIGURE 6.2
Capture zone of the extraction well (Example 6.5).

2. Do not specify the y values beyond the values of $\pm Q/2Bu$. As discussed, $\pm Q/2Bu$ are the asymptotic values of the capture zone curve ($x = \infty$).

Example 6.6: Determine the Downstream and Sidestream Distances of a Capture Zone

A groundwater extraction well is installed in an aquifer (hydraulic conductivity $= 1,000$ gpd/ft², hydraulic gradient $= 0.015$, and aquifer thickness $= 80$ ft). The design pumping rate is 50 gpm. Delineate the capture zone of this recovery well by specifying the following characteristic distances of the capture zone:

(a) The sidestream distance from the well to the envelope of the capture zone at the line of the pumping well

(b) The downstream distance from the well to stagnation point of the envelope

(c) The sidestream distance of the envelope far upstream of the pumping well

Solution:

(a) Determine the groundwater velocity, u:

$$u = (K)(i) = [(1,000 \text{ gal/d/ft}^2)(1 \text{ d/1,440 min})(1 \text{ ft}^3/7.48 \text{ gal})](0.015)$$
$$= 1.39 \times 10^{-3} \text{ ft/min}$$

(b) Determine the sidestream distance from the well to the envelope of the capture zone at the line of the pumping well:

$$\frac{Q}{4Bu} = \frac{(50 \text{ gal/min})(1 \text{ ft}^3/7.48 \text{ gal})}{(4)(80 \text{ ft})(1.39 \times 10^{-3} \text{ ft/min})} = 15.0 \text{ ft}$$

(c) Determine the downstream distance from the well to stagnation point of the envelope:

$$\frac{Q}{2\pi Bu} = \frac{(50 \text{ gal/min})(1 \text{ ft}^3/7.48 \text{ gal})}{(2)(\pi)(80 \text{ ft})(1.39 \times 10^{-3} \text{ ft/min})} = 9.6 \text{ ft}$$

(d) Determine the sidestream distance of the envelope far upstream of the pumping well:

$$\frac{Q}{2Bu} = \frac{(50 \text{ gal/min})(1 \text{ ft}^3/7.48 \text{ gal})}{(2)(80 \text{ ft})(1.39 \times 10^{-3} \text{ ft/min})} = 30.0 \text{ ft}$$

(e) The general shape of the envelope can be defined by using the given characteristic distances:

x (ft)	y (ft)	Note
0	0	well location
−9.6	0	downstream distance (stagnation point)
0	15	sidestream distance at the line of the well
0	−15	sidestream distance at the line of the well
150*	30	sidestream distance at far upstream of the well
150*	−30	sidestream distance at far upstream of the well

*The sidestream distance at far upstream of the well, ± 30 ft, should occur at x = ∞. A value of 150, which is ten times the sidestream distance at the line of well, is used here as the value of x.

(f) The data in this table are plotted in Figure 6.3 (the solid squares). The envelopes generated using Equations (6.4) and (6.5) are also plotted for comparison.

6.2.2.2 Multiple Wells

Table 6.1 tabulates some characteristic distances of the capture zone for a network of groundwater extraction wells located on a line perpendicular to the flow direction. As shown in the table, the distance between the dividing streamlines far upstream from the pumping wells is equal to $n(Q/Bu)$, where

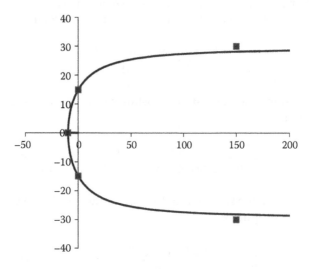

FIGURE 6.3
Capture zone of the extraction well (Example 6.6).

TABLE 6.1

Characteristic Distances of the Capture Zone of a Network of Pumping Wells

No. of Extraction Wells	Optimal Distance between Each Pair of Extraction Wells	Distance between the Streamlines at the Line of the Wells	Distance between the Streamlines at Far Upstream from the Wells
1	...	$0.5(Q/Bu)$	(Q/Bu)
2	$0.32(Q/Bu)$	(Q/Bu)	$2(Q/Bu)$
3	$0.40(Q/Bu)$	$1.5(Q/Bu)$	$3(Q/Bu)$
4	$0.38(Q/Bu)$	$2(Q/Bu)$	$4(Q/Bu)$

Source: Modified from [1].

n is the number of the pumping wells. This distance is twice the distance between the streamlines at the line of the wells.

The downstream distance of a network of wells is very similar to that of one pumping well, which is equal to $Q/2\pi Bu$.

Example 6.7: Determine the Downstream and Sidestream Distances of the Capture Zone of a Network of Wells

Two groundwater extraction wells are to be installed in an aquifer (hydraulic conductivity = 1,000 gpd/ft², hydraulic gradient = 0.015, and aquifer thickness = 80 ft).

The design pumping rate of each well is 50 gpm. Determine the optimal distance between these two wells and delineate the capture zone of these extraction wells by specifying the following characteristic distances of the capture zone (see Figure 6.4):

(a) The sidestream distance from the wells to the envelope of the capture zone at the line of the pumping wells

(b) The downstream distance from the wells to stagnation points of the envelope

(c) The sidestream distance of the envelope far upstream of the pumping wells

Solution:

(a) Determine the groundwater velocity, u:

$u = (K)(i) = [(1{,}000 \text{ gal/d/ft}^2)(1 \text{ d}/1{,}440 \text{ min})(1 \text{ ft}^3/7.48 \text{ gal})](0.015)$
   $= 1.39 \times 10^{-3}$ ft/min

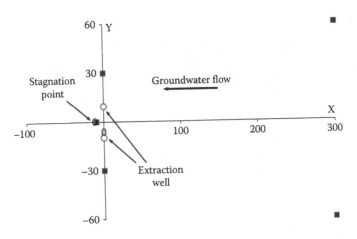

FIGURE 6.4
Capture zone of two extraction wells (Example 6.7).

(b) Determine the optimum distance between these two wells, which is $0.32Q/Bu$ (from Table 6.1):

$$\frac{0.32Q}{Bu} = \frac{(0.32)(50 \text{ gal / min})(1 \text{ ft}^3/7.48 \text{ gal})}{(80 \text{ ft})(1.39 \times 10^{-3} \text{ ft/min})} = 19.2 \text{ ft}$$

The distance of each well to the origin is half of this value $= 0.16Q/Bu = 9.6$ ft.

(c) Determine the sidestream distance from the well to the envelope of the capture zone at the line of the pumping well, $Q/2Bu$:

$$\frac{Q}{2Bu} = \frac{(50 \text{ gal / min})(1 \text{ ft}^3/7.48 \text{ gal})}{(2)(80 \text{ ft})(1.39 \times 10^{-3} \text{ ft / min})} = 30.0 \text{ ft}$$

(d) Determine the downstream distance from the well to stagnation point of the envelope, $Q/2\pi Bu$:

$$\frac{Q}{2\pi Bu} = \frac{(50 \text{ gal/min})(1 \text{ ft}^3/7.48 \text{ gal})}{(2)(\pi)(80 \text{ ft})(1.39 \times 10^{-3} \text{ ft/min})} = 9.6 \text{ ft}$$

(e) Determine the sidestream distance of the envelope far upstream of the pumping wells, Q/Bu:

$$\frac{Q}{Bu} = \frac{(50 \text{ gal/min})(1 \text{ ft}^3/7.48 \text{ gal})}{(80 \text{ ft})(1.39 \times 10^{-3} \text{ ft/min})} = 60.0 \text{ ft}$$

(f) The general shape of the envelope can be defined by using the above characteristic distances:

x (ft)	y (ft)	Note
0	9.6	location of the first well
0	-9.6	location of the second well
-9.6	0	downstream distance (stagnation point)
0	30	sidestream distance at the line of the wells
0	-30	sidestream distance at the line of the wells
300*	60	sidestream distance far upstream of the wells
300*	-60	sidestream distance far upstream of the wells

*The sidestream distance far upstream of the wells, ±60 ft, should occur at $x = \infty$. A value of 300, which is 10 times the sidestream distance at the line of wells, is used as the value of x.

Discussion:

1. The sidestream distance at the line of the two pumping wells is twice that of one well.

2. The sidestream distance far upstream of the two pumping wells is twice that of one well.

3. The downstream distance of the two pumping wells is the same as that of one pumping well. The calculated downstream distance, $Q/2\pi Bu$, is along the x-axis. However, the affected distances directly downstream of these two wells should be slightly larger than $Q/2\pi Bu$.

6.2.2.3 Well Spacing and Number of Wells

As mentioned previously, it is important to determine the number of wells and their spacing in a groundwater remediation program. After the extent of the plume as well as the direction and velocity of the groundwater flow have been determined, the following procedure can be used to determine the number of wells and their locations:

Step 1: Determine the groundwater pumping rate from aquifer testing or estimate the flow rate by using information of the aquifer materials.

Step 2: Draw the capture zone of one groundwater well (see Example 6.5 or 6.6), using the same scale as the plume map.

Step 3: Superimpose the capture-zone curve on the plume map.

Step 4: If the capture zone can completely encompass the extent of the plume, one pumping well is sufficient. The location of the well on the capture-zone curve is then copied to the plume map. One may want to reduce the groundwater extraction rate to have a smaller capture zone, but still sufficient to cover the entire plume.

Step 5: If the capture zone cannot encompass the entire plume, pre- pare the capture-zone curves using two or more pumping wells until the capture zone can cover the entire plume. The well locations on the capture-zone curve are then copied to the plume map. Note that the zones of influence of individual wells may overlap. One may not be able to pump the same flow rate from each well in a network of wells as one can from a single well with the same allowable drawdown.

Example 6.8: Determine the Number and Locations of Pumping Wells to Capture a Groundwater Plume

An aquifer (hydraulic conductivity = 1,000 gpd/ft^2, gradient = 0.015, and aquifer thickness = 80 ft) is impacted. The extent of the plume has been defined and shown in Figure 6.5.

Determine the number and locations of groundwater extraction wells for remediation. The design pumping rate of each well is 50 gpm.

Solution:

(a) Plot the capture zone of one extraction well (same as Example 6.6), and locate the well at the origin of the coordinate system. The dotted envelopes on the figure define the capture zone of this well. As shown, this capture zone could not encompass the entire plume.

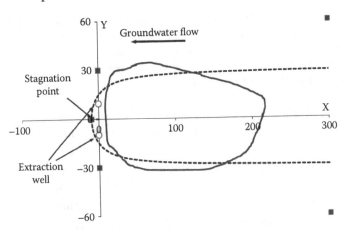

FIGURE 6.5
The plume and the capture zones of one well and two extraction wells (Example 6.8).

(b) Plot the capture zone of two pumping wells (same as Example 6.7). The two open circles on Figure 6.5 are the locations of the wells, and the square symbols define the capture zone of these two wells. As shown, this capture zone cannot encompass the entire plume. Consequently, two pumping wells should be employed.

6.3 Activated-Carbon Adsorption

6.3.1 Description of the Activated-Carbon Adsorption

Adsorption is the process that collects soluble substances (adsorbates) in solution onto the surface of the solid (the adsorbent). Activated carbon is a universal adsorbent that adsorbs almost all types of organic compounds. Granular or powder activated carbon has a large specific surface area. In activated-carbon adsorption, the organics leave the liquid by adsorbing onto the activated-carbon surface. As the activated-carbon unit becomes exhausted, as indicated by the breakthrough of COCs in the effluent, the activated-carbon unit needs to be regenerated or replaced.

Common preliminary design of an activated-carbon adsorption system includes sizing the adsorption unit, determining the carbon-change (or regeneration) interval, and configuring the carbon units, if multiple adsorption units are used.

6.3.2 Adsorption Isotherm and Adsorption Capacity

In general, the extent of adsorption depends on the characteristics of the adsorbates (i.e., COCs) and the activated carbon, concentrations of the COCs, and the temperature. An adsorption isotherm describes the equilibrium relationship between the adsorbed COC concentration on the surface of the activated carbon and the dissolved COC concentration in the bulk solution at a given temperature. The adsorption capacity of a given activated carbon for a specific compound is estimated from its isotherm data. The most commonly used adsorption models in environmental applications are the Langmuir and Freundlich isotherms, respectively:

$$q = \frac{abC}{1+bC} \tag{6.6}$$

$$q = kC^n \tag{6.7}$$

where q is the adsorbed COC concentration (in mass of COC/mass of activated carbon), C is the aqueous COC concentration (in mass of COC/volume of the solution), and a, b, k, and n are constants. The adsorbed COC concentration (q) obtained from Equations (6.6) or (6.7) is an equilibrium value (the one in equilibrium with the aqueous COC concentration). It should be considered as the theoretical adsorption capacity for a specified aqueous COC concentration. The actual adsorption capacity in the field applications should be lower because the adsorption isotherms are usually developed in a laboratory setting, where other compounds that would compete for the adsorption sites are absent. Normally, design engineers take 25% to 50% of this theoretical value as the design adsorption capacity as a factor of safety. Therefore,

$$q_{design} = (50\%)(q_{theoretical}) \qquad (6.8)$$

The maximum amount of COCs that can be removed from water and held by a given amount of activated carbon ($M_{removal}$) can be determined as:

$$M_{removal} = (q_{design})(M_{activated\ carbon})$$

$$= (q_{design})[(V_{activated\ carbon})(\rho_b)] \qquad (6.9)$$

where $M_{activated\ carbon}$ is the mass, $V_{activated\ carbon}$ is the volume, and ρ_b is the bulk density of the activated carbon, respectively.

The following procedure can be used to determine the adsorption capacity of an activated-carbon adsorption unit (often called an *adsorber*):

Step 1: Determine the theoretical adsorption capacity of the activated carbon by using Equations (6.6) or (6.7).

Step 2: Determine the design adsorption capacity of the activated carbon by using Equation (6.8).

Step 3: Determine the amount of activated carbon in the adsorber.

Step 4: Determine the maximum amount of COCs that can be held by the activated carbon in the adsorber using Equation (6.9).

Information needed for this calculation:

- The adsorption isotherm
- COC concentration of the influent liquid, C_{in}
- Volume of the activated carbon, $V_{activated\ carbon}$
- Bulk density of the activated carbon, ρ_b

Example 6.9: Determine the Capacity of an Activated-Carbon Adsorber

Dewatering to lower the groundwater level for belowground construction is often necessary. At a construction site, the contractor unexpectedly found that the extracted groundwater contained 5 mg/L toluene. The toluene concentration of the groundwater has to be reduced to below 100 ppb before discharge. To avoid further delay of the tight construction schedule, off-the-shelf 55-gallon activated-carbon units are proposed to treat the extracted groundwater.

The activated-carbon vendor provided the adsorption isotherm information. It follows the Langmuir model as: q(kg toluene/kg carbon) = $[0.04C_e/(1+0.002C_e)]$ where C_e is in mg/L. The vendor also provided the following information regarding the adsorber:

- Diameter of carbon packing bed in each 55-gal drum = 1.5 ft
- Height of carbon packing bed in each 55-gal drum = 3 ft
- Bulk density of the activated carbon = 30 lb/ft³

Determine (1) the adsorption capacity of the activated carbon, (2) the amount of activated carbon in each 55-gallon unit, and (3) the amount of the toluene that each unit can remove before becoming exhausted.

Solution:

(a) The theoretical adsorption capacity can be found by using the given adsorption isotherm as:

$$q(\text{kg/kg}) = \frac{0.004C_e}{1+0.002C_e} = \frac{(0.004)(5)}{1+(0.002)(5)} = 0.02 \text{ kg/kg}$$

The actual adsorption capacity can be found by using Equation (6.8) as:

$q_{design} = (50\%)q_{theoretical} = (50\%)(0.02) = 0.01$ kg toluene/kg activated carbon

(b) Volume of the activated carbon inside a 55-gal drum

$= (\pi r^2)(h)$

$= (\pi)[(1.5/2)^2](3) = 5.3$ ft³

Amount of the activated carbon inside a 55-gal drum = $(V)(\rho_b)$

$= (5.3 \text{ ft}^3)(30 \text{ lb/ft}^3) = 159$ lb

(c) Amount of toluene that can be retained in a drum before the carbon becomes exhausted

$=$ (amount of the activated carbon)(actual adsorption capacity)
$= (159 \text{ lb/drum})(0.01 \text{ lb toluene/lb activated carbon}) = 1.59$ lb toluene/drum

Discussion:

1. The bulk density of activated carbon is typically in the neighborhood of 30 lb/ft³. The amount of activated carbon in a 55-gallon drum is approximately 160 lb.

2. The adsorption capacity of 0.01 kg/kg is equal to 0.01 lb/lb or 0.01 g/g.

3. Care should be taken to use matching units for C and q in the isotherm equations.

4. The influent aqueous COC concentration, not the effluent concentration, should be used in the isotherms to estimate the adsorption capacity.

6.3.3 Design of an Activated-Carbon Adsorption System

6.3.3.1 Empty-Bed Contact Time

To size the liquid-phase activated-carbon system, the common criterion used in design is the empty-bed contact time (EBCT). The typical EBCT ranges from 5 to 20 min, mainly depending on the characteristics of the COCs. Some compounds have a stronger tendency to adsorption, and the required EBCT would be shorter. Taking PCBs (polychlorinated biphenyls) and acetone as two extreme examples, PCBs are very hydrophobic and will strongly adsorb to the activated-carbon surface, while acetone is very soluble in water and not readily adsorbable. The required EBCT for acetone would be much longer than that for PCBs.

If the liquid flow rate (Q) is specified, the EBCT can be used to determine the required volume of the activated-carbon adsorber ($V_{\text{activated carbon}}$) as:

$$V_{\text{activated carbon}} = (Q)(EBCT) \tag{6.10}$$

6.3.3.2 Cross-Sectional Area

The typical hydraulic loading rate to carbon adsorbers is set to be ≤5 gpm/ft². This parameter can be used to determine the minimum required cross-sectional area of the adsorber ($A_{\text{activated carbon}}$):

$$A_{\text{activated carbon}} = \frac{Q}{\text{surface loading rate}} \tag{6.11}$$

6.3.3.3 Height of the Activated-Carbon Adsorber

The required height of the activated-carbon adsorber ($H_{\text{activated carbon}}$) can then be determined as:

$$H_{\text{activated carbon}} = \frac{V_{\text{activated carbon}}}{A_{\text{activated carbon}}} \tag{6.12}$$

6.3.3.4 COC Removal Rate by the Activated-Carbon Adsorber

The removal rate by an activated-carbon adsorber ($R_{removal}$) can be calculated by using the following formula:

$$R_{removal} = (C_{in} - C_{out})Q \tag{6.13}$$

In practical applications, the effluent concentration (C_{out}) is kept below the discharge limit, which is often very low. Therefore, for a factor of safety, the term of C_{out} can be deleted from Equation (6.13) in design. The mass removal rate is then essentially the same as the mass loading rate ($R_{loading}$):

$$R_{removal} \cong R_{loading} = (C_{in})Q \tag{6.14}$$

6.3.3.5 Change-Out (or Regeneration) Frequency

Once the activated carbon reaches its capacity, it should be regenerated or disposed of. The time interval between two consecutive regenerations or the expected service life of a fresh batch of activated carbon can be estimated by dividing the capacity of the activated carbon in the adsorber with the COC removal rate ($R_{removal}$) as:

$$T = \frac{M_{removal}}{R_{removal}} \tag{6.15}$$

6.3.3.6 Configuration of the Activated-Carbon Adsorbers

If multiple activated-carbon adsorbers are used, the adsorbers are often arranged in series and/or in parallel. If two adsorbers are arranged in series, a monitoring point should be located at the effluent of the first adsorber. A high effluent concentration from the first adsorber indicates that this adsorber is reaching its capacity. The first adsorber is then taken off-line and the second adsorber is shifted to be the first adsorber. Consequently, the capacity of both adsorbers would be fully utilized, and the compliance requirements are met. If there are two parallel streams of adsorbers, one stream can always be taken off-line for regeneration or maintenance, and the continuous operation of the process will not be interrupted.

The following procedure can be used to complete the design of an activated-carbon adsorption system:

Step 1: Determine the design adsorption capacity as described previously (also see Example 6.9).

Step 2: Determine the required volume of the activated-carbon adsorber by using Equation (6.10).

Step 3: Determine the required area of the activated-carbon adsorber by using Equation (6.11).

Step 4: Determine the required height of the activated-carbon adsorber by using Equation (6.12).

Step 5: Determine the COC removal rate or loading rate by using Equation (6.14).

Step 6: Determine the amount of the COCs that the carbon adsorber(s) can hold by using Equation (6.9).

Step 7: Determine the service life of the activated-carbon adsorber by using Equation (6.15).

Step 8: Determine the optimal configuration when multiple adsorbers are used.

Information needed for this calculation:

- The adsorption isotherm
- COC concentration in the influent liquid, C_{in}
- Design hydraulic loading rate
- Design liquid flow rate, Q
- Bulk density of the activated carbon, ρ_b

Example 6.10: Design an Activated-Carbon System for Groundwater Remediation

Dewatering to lower the groundwater level for belowground construction is often necessary. At a construction site, the contractor unexpectedly found that the extracted groundwater contained 5 mg/L toluene. The toluene concentration of the groundwater has to be reduced to below 100 ppb before discharge. To avoid further delay of the tight construction schedule, off-the-shelf 55-gallon activated-carbon units are proposed to treat the groundwater. Use the following information to design an activated-carbon treatment system (i.e., number of carbon units, configuration of flow, and carbon change-out frequency):

- Wastewater flow rate = 30 gpm
- Diameter of carbon packing bed in each 55-gal drum = 1.5 ft
- Height of carbon packing bed in each 55-gal drum = 3 ft
- Bulk density of GAC = 30 lb/ft^3
- Adsorption isotherm: q(kg toluene/kg carbon) = $[0.04C_e/(1+0.002C_e)]$, where C_e is in mg/L

Solution:

(a) The design adsorption capacity has been found to be 0.01 lb/lb in Example 6.9.

(b) Assuming an EBCT of 12 min, the required volume of the activated-carbon adsorber can be found by using Equation (6.10):

$$V_{carbon} = (Q)(EBCT) = [(30 \text{ gpm})(ft^3/7.48 \text{ gal})](12 \text{ min}) = 48.1 \text{ ft}^3$$

(c) Assuming a design hydraulic loading of 5 gpm/ft² or less, the required cross-sectional area for carbon adsorption can be found by using Equation (6.11):

$$A_{carbon} = \frac{Q}{\text{surface loading rate}} = \frac{30 \text{ gpm}}{5 \text{ gpm/ft}^2} = 6 \text{ ft}^2$$

(d) If the adsorption system is tailor-made, then a system with a cross-sectional area of 6 ft² and a height of 8 ft (= 48.1/6) will do the job. However, because the off-the-shelf 55-gallon drums are to be used, we need to determine the number of the drums that will provide the required cross-sectional area.

Area of the activated carbon inside a 55-gal drum = (πr^2)

$$= (\pi)[(1.5/2)^2] = 1.77 \text{ ft}^2/\text{drum}$$

Number of drums in parallel to meet the required hydraulic loading rate

$$= (6 \text{ ft}^2) \div (1.77 \text{ ft}^2/\text{drum}) = 3.4 \text{ drums}$$

So, use four drums in parallel. The total cross-sectional area of four drums is equal to 7.08 ft² (= 1.77 × 4).

(e) The required height of the activated-carbon adsorber can be found by using Equation (6.12):

$$H_{carbon} = \frac{V_{carbon}}{A_{carbon}} = \frac{48.1}{7.08} = 6.8 \text{ ft}$$

The height of activated carbon in each drum is 3 ft. The number of drums required in series to meet the required height of 6.8 ft can be found as:

Number of drums in series to meet the required height

$$= (6.8 \text{ ft}) \div (3 \text{ ft/drum}) = 2.3 \text{ drums}$$

So, use three drums in series for each of the four process trains. The total volume of activated carbon in 12 drums is equal to 63.6 ft³ (= 5.3 ft³/drum × 12 drums).

(f)   Determine the toluene removal rate or loading rate by using Equation (6.14):

$$R_{removal} \approx R_{loading} = (C_{in})Q$$

$$= (5 \text{ mg/L})(30 \text{ gal/min})(3.785 \text{ L/gal})(1,440 \text{ min/day})]$$

$$= 817,560 \text{ mg/day} = 1.8 \text{ lb/day}$$

(g)   Determine the amount of toluene that the carbon adsorber(s) can hold by using Equation (6.9):

$$M_{removal} = (q_{actual})[(V_{carbon})(\rho_b)]$$

$$= (0.01)[(63.6)(30)] = 19.1 \text{ lb}$$

(h)   Determine the service life of the carbon adsorbers by using Equation (6.15):

$$T = \frac{M_{removal}}{R_{removal}} = \frac{19.1 \text{ lb}}{1.8 \text{ lb/d}} = 10.6 \text{ days}$$

Discussion:

1. The configuration is four process trains in parallel with three drums in series for each train (a total of 12 drums). Care should be taken to minimize the head loss resulting from three drums in series and numerous piping connections.

2. A 55-gallon activated-carbon drum normally costs several hundred dollars. In this example, 12 drums last less than 11 days. The disposal or regeneration cost should also be added, and it makes this option relatively expensive. If a long-term treatment is needed, one may want to switch to larger activated-carbon adsorbers or to other treatment methods.

6.4 Air Stripping

6.4.1 Description of the Air-Stripping Process

Air stripping is a physical process that enhances volatilization of organic compounds from water by passing clean air through it. It is one of the commonly used processes for treating groundwater impacted by volatile organic compounds (VOCs).

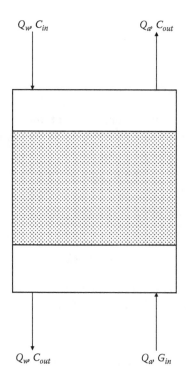

FIGURE 6.6
Schematic diagram of a packed-column air stripper.

An air-stripping system creates air and water interfaces to enhance mass transfer between the air and liquid phases. Although there are several system configurations commercially available, including tray columns, spray systems, diffused aeration, and packed columns (or often called *packed towers*), use of packed towers is the most popular alternative for groundwater remediation applications (Figure 6.6).

6.4.2 Design of an Air-Stripping System

In a packed-column air-stripping tower, the air and the impacted groundwater streams flow countercurrently through a packing column. The packing provides a large surface area for VOCs to migrate from the liquid stream to the air stream. A mass-balance equation can be derived by letting the amount of COCs removed from the liquid be equal to the amount of the COCs that enter the air:

$$Q_w(C_{in} - C_{out}) = Q_a(G_{out} - G_{in}) \qquad (6.16)$$

where
   $C =$ COC concentration in the liquid phase (mg/L)
   $G =$ COC concentration in the air phase (mg/L)
   $Q_a =$ air flow rate (L/min)
   $Q_w =$ liquid flow rate (L/min)

For an ideal case where the influent air contains no COCs ($G_{in} = 0$) and the groundwater is completely decontaminated ($C_{out} = 0$), Equation (6.16) can be simplified as:

$$Q_w(C_{in}) = Q_a(G_{out}) \tag{6.17}$$

Assume that Henry's law applies and the effluent air is in equilibrium with the influent water; then:

$$G_{out} = H^*C_{in} \tag{6.18}$$

where H^* is the Henry's constant of the COC in a dimensionless form.
   Combining Equations (6.17) and (6.18), the following relationship can be developed:

$$H^*\left(\frac{Q_a}{Q_w}\right)_{min} = 1 \tag{6.19}$$

The $(Q_a/Q_w)_{min}$ is the minimum air-to-water ratio (in vol/vol), and this is the air-to-water ratio for the previously mentioned ideal case. The actual air-to-water ratio is often chosen to be a few times larger than the minimum air-to-water ratio.
   The stripping factor (S), which is the product of the dimensionless Henry's constant and the air-to-water ratio, is commonly used in design:

$$S = H^*\left(\frac{Q_a}{Q_w}\right) \tag{6.20}$$

The stripping factor is equal to unity for the previously mentioned ideal case. It would require a packing height of infinity to achieve the perfect removal. For field applications, the values of S should be greater than 1. Practical values of S range from 2 to 10. Operating the system with a value of S larger than 10 may not be economical. In addition, a high air-to-water ratio may cause an unfavorable phenomenon, called *flooding*, in air-stripping operations.
   The following procedure can be used to determine the air flow rate for a given liquid flow rate:

   Step 1: Convert the Henry's constant to its dimensionless value using
         the formula given in Table 2.4.

Step 2: If the stripping factor is known or selected, determine the air-to-water ratio by using Equation (6.20). Go to Step 4.

Step 3: If the stripping factor is not known or selected, determine the minimum air-to-water ratio by using Equation (6.19). Obtain the design air-to-water ratio by multiplying this minimum air-to-water ratio with a value between 2 to 10. Go to Step 4.

Step 4: Determine the required air flow rate by multiplying the liquid flow rate with the air-to-water ratio determined from Step 2 or Step 3.

Information needed for this calculation:

- Henry's constant
- Stripping factor, S
- Design liquid flow rate, Q

Example 6.11: Determine the Air-to-Water Ratio for an Air Stripper

A packed-column air stripper is designed to reduce the chloroform concentration in the extracted groundwater. The concentration is to be reduced from 50 mg/L to 0.05 mg/L (50 ppb). Determine (1) the minimum air-to-water ratio, (2) the design air-to-water ratio, and (3) the design air flow rate. Use the following information in calculations:

- Henry's constant for chloroform = 128 atm
- Stripping factor = 3
- Temperature of the water = 15°C
- Extracted groundwater flow rate = 120 gpm

Solution:

(a) Use the formula in Table 2.4 to convert the Henry's constant to its dimensionless value:

$$H = \frac{H^* RT(1000\gamma)}{W} = \frac{H^*(0.082)(273+15)(1000)(1)}{18} = 128$$

So, $H^* = 0.098$ (dimensionless)

(b) Use Equation (6.19) to determine the minimum air-to-water-ratio:

$$H^*\left(\frac{Q_a}{Q_w}\right)_{min} = 1 = (0.098)\left(\frac{Q_a}{Q_w}\right)_{min}$$

So, $(Q_a/Q_w)_{min} = 10.25$ (dimensionless)

(c) Use Equation (6.20) to determine the air-to-water-ratio:

$$H^* \left(\frac{Q_a}{Q_w} \right) = S = 3 = (0.098) \left(\frac{Q_a}{Q_w} \right)$$

So, $(Q_a/Q_w) = 30.75$ (dimensionless)

(d) Determine the required air flow rate by multiplying the liquid flow rate with the air-to-water ratio:

$Q_a = Q_w \times (Q_a/Q_w) = (120 \text{ gpm})(30.75) = 3,690 \text{ gal/min} = 493 \text{ ft}^3/\text{min}$

Discussion:

A stripping factor of three means that the ratio of the design Q_a/Q_w and the minimum Q_a/Q_w is also three. Consequently, the design Q_a/Q_w can be obtained by multiplying the minimum Q_a/Q_w with the stripping factor.

6.4.2.1 Column Diameter

One of the key components in sizing an air stripper is to determine the diameter of the column. The diameter depends mainly on the liquid flow rate. The higher the liquid flow rate is, the larger the column diameter would be. Typical liquid hydraulic loading rate to an air-stripping column is kept to 20 gpm/ft² or less. This parameter is often used to determine the required cross-sectional area of the stripping column ($A_{stripping}$):

$$A_{stripping} = \frac{Q}{\text{surface loading rate}} \qquad (6.21)$$

6.4.2.2 Packing Height

The required depth of the packing column (Z) for a specific removal efficiency is another important design component. A taller column would be required to achieve a larger removal efficiency. The packing height can be determined using the transfer unit concept:

$$Z = (HTU) \times (NTU) \qquad (6.22)$$

where HTU is the height of transfer unit and NTU is the number of transfer units.

The HTU value depends on the hydraulic loading rate and the overall liquid-phase mass-transfer coefficient, $K_L a$. (Note: K_L is the rate constant [m/s] and a is the specific surface area [m²/m³]. $K_L a$ has a unit of 1/time.) The $K_L a$ value for a specific application can be best determined from pilot testing, and there are also empirical equations available to estimate the value of $K_L a$. Values of $K_L a$ in common air-stripping columns used in groundwater remediation range from 0.01 to 0.05 s^{-1}. HTU has a unit of length and can be determined as:

$$HTU = \frac{Q_L}{(K_L a)} \tag{6.23}$$

where Q_L is the hydraulic loading rate in length/time.

The NTU value can be determined by using the following formula:

$$NTU = \left(\frac{S}{S-1}\right) \ln\left\{\frac{(C_{in} - G_{in}/H^*)}{(C_{out} - G_{in}/H^*)}\left(\frac{S-1}{S}\right) + \frac{1}{S}\right\}$$

$$= \left(\frac{S}{S-1}\right) \ln\left\{\left(\frac{C_{in}}{C_{out}}\right)\left(\frac{S-1}{S}\right) + \frac{1}{S}\right\} \quad (\text{for } G_{in} = 0) \tag{6.24}$$

where S is the stripping factor, H^* is the Henry's constant in dimensionless form, C is the COC concentration in liquid, and G is the COC concentration in air.

The following procedure can be used to size an air-stripping column:

Step 1: Determine the required cross-sectional area of the air stripper by using Equation (6.21). Then, determine the diameter of the column corresponding to this calculated area. Round up the diameter value to the next half or whole foot.

Step 2: Use the newly found diameter to calculate the cross-sectional area and, then, the hydraulic loading rate. Use Equation (6.23) to find the HTU value.

Step 3: Determine the stripping factor, if not known or specified, by using Equation (6.20).

Step 4: Use Equation (6.24) to find the NTU value.

Step 5: Use Equation (6.22) to find the packing height, Z.

Information needed for this calculation:

- Henry's constant
- Stripping factor, S
- Design hydraulic loading rate
- Design liquid flow rate, Q_L
- Overall liquid-phase mass-transfer coefficient, $K_L a$
- COC concentration in the influent liquid, C_{in}

- COC concentration in the effluent liquid, C_{out}
- COC concentration in influent air, G_{in}

Example 6.12: Sizing an Air Stripper for Groundwater Remediation

A packed-column air stripper is designed to reduce chloroform concentration in the extracted groundwater. The concentration is to be reduced from 50 mg/L to 0.05 mg/L (50 ppb). Size the air stripper by determining the air flow rate, cross-sectional surface area, and packing height.

Use the following information in calculations:

- Henry's constant for chloroform = 128 atm
- Stripping factor = 3
- Temperature of the water = 15°C
- Extracted groundwater flow rate = 120 gpm
- $K_L a = 0.01/s$
- Type of packing = Jaeger 3" Tri-packs
- Hydraulic loading rate = 20 gpm/ft²
- Chloroform concentration in the influent air = 0

Solution:

(a) As shown in Example 6.11, the dimensionless value of the Henry's constant is equal to 0.098, and the air flow rate has been determined to be 493 ft³/min.

(b) Use Equation (6.19) to determine the required cross-sectional area:

$$A_{stripping} = \frac{Q}{\text{surface loading rate}} = \frac{120 \text{ gpm}}{20 \text{ gpm/ft}^2} = 6 \text{ ft}^2$$

Diameter of the air stripping column $= (4 \times A/\pi)^{1/2}$
$= (4 \times 6/\pi)^{1/2} = 2.76 \text{ ft}$

So, $d = 3$ ft

(c) Use this newly found diameter to find the hydraulic loading rate:

Cross-sectional area of the column $= \pi d^2/4 = \pi (3)^2/4 = 7.1 \text{ ft}^2$

Hydraulic loading rate to the column $(Q_L) = Q/A$
$= [(120 \text{ gpm})(\text{ft}^3/7.48 \text{ gal})] \div 7.1 \text{ ft}^2 = 2.26 \text{ ft/min} = 0.0377 \text{ ft/s}$

(d) Use Equation (6.23) to determine the HTU value:

$$HTU = \frac{L}{(K_L a)} = \frac{0.0377 \text{ ft/s}}{0.01/s} = 3.77 \text{ ft}$$

(e) Use Equation (6.24) to determine the NTU value:

$$NTU = \left(\frac{S}{S-1}\right) \ln\left\{\left(\frac{C_{in}}{C_{out}}\right)\left[\frac{S-1}{S}\right] + \frac{1}{S}\right\} \quad (\text{for } G_{in} = 0)$$

$$= \left(\frac{3}{3-1}\right) \ln\left\{\left(\frac{50}{0.05}\right)\left[\frac{3-1}{3}\right] + \frac{1}{3}\right\} = 9.75$$

(f) Use Equation (6.22) to determine the packing height:

$$Z = (HTU) \times (NTU) = (3.77 \text{ ft})(9.75) = 36.8 \text{ ft}$$

Discussion:

1. The typical hydraulic loading rate, 20 gpm/ft², is much higher than that for the activated-carbon adsorbers, 5 gpm/ft².
2. The required packing height of 36.8 ft will make the total height of the air stripper well over 40 feet. This may not be acceptable in most project locations. If this is the case, one may consider having two shorter air strippers in series.

6.5 *Ex Situ* Biological Treatment

6.5.1 Description of the *Ex Situ* Biological Treatment Process

Biological processes can be used to remove biodegradable organic compounds from water. With regard to groundwater remediation, the impacted groundwater can be treated biologically in the *in situ* or the *ex situ* mode. This section covers the *ex situ* biological treatment, while the *in situ* bioremediation of the impacted aquifer will be covered in the next section. Aboveground biological reactors can be employed to remove organic COCs from the extracted groundwater. In general, the bioreactors for removal of dissolved organics from water/wastewater can be classified into two groups: suspended growth and attached growth. The most common type of suspended growth is the activated-sludge process, while that for the attached growth is the trickling-filter process.

Biological systems used in groundwater remediation are usually much smaller in scale compared to those in municipal or industrial wastewater treatment plants. The trickling filters consist of packing materials to support the bacterial growth. Since the biological process is relatively complicated and affected by many factors, a feasibility study as well as bench- and/or pilot-scale testing are recommended before the process is seriously considered as a viable remedial alternative.

6.5.2 Design of an Aboveground Biological System

For the trickling-filter type of bioreactors, the following empirical equation is often used [2]:

$$\frac{C_{out}}{C_{in}} = \exp\left[-kD(Q/A)^{-0.5}\right] \tag{6.25}$$

where
C_{out} = COC concentration in the reactor effluent, mg/L
C_{in} = COC concentration in the reactor influent, mg/L
k = rate constant corresponding to the packing depth of D, $(gpm)^{0.5}$/ft
D = depth of the filter, ft
Q = liquid flow rate, gpm
A = cross-sectional area of the packing material, ft^2

The hydraulic loading rate to a bioreactor is often small, at 0.5 gpm/ft^2 or less. If the hydraulic loading rate is known, Equation (6.27) can be used to determine the cross-sectional area of the bioreactor:

$$A_{bioreactor} = \frac{Q}{\text{surface loading rate}} \tag{6.26}$$

When a rate constant determined from a different packing depth is used, the following empirical formula should be applied to adjust the rate constant:

$$k_2 = k_1\left(\frac{D_1}{D_2}\right)^{0.3} \tag{6.27}$$

where
k_1 = rate constant corresponding to a filter of depth D_1
k_2 = rate constant corresponding to a filter of depth D_2
D_1 = depth of filter #1
D_2 = depth of filter #2

The following procedure can be used to size an attached-growth bioreactor:

Step 1: Select a desirable packing depth, D. Adjust the rate constant to the selected packing depth, if necessary, by using Equation (6.27).

Step 2: Determine the hydraulic loading rate of the bioreactor by using Equation (6.25).

Step 3: Determine the required cross-sectional area by using Equation (6.26). Calculate the diameter of the bioreactor corresponding to this area. Round up the diameter value to the next half or whole foot. If the calculated cross-sectional area is too large, select a larger packing depth and restart from Step 1.

Information needed for this calculation:

- Rate constant, k
- Influent COC concentration, C_{in}
- Effluent COC concentration, C_{out}
- Design liquid flow rate, Q

Example 6.13: Sizing an Aboveground Bioreactor for Groundwater Remediation

A packed-bed bioreactor is designed to reduce the toluene concentration in the extracted groundwater from 4 to 0.1 mg/L (100 ppb). The packing depth has been selected to be 3 ft. Determine the required diameter of the bioreactor.

Use the following information in calculations:

- Rate constant = 0.9 $(gpm)^{0.5}$/ft at 20°C (for 2-ft packing depth)
- Temperature of the water = 20°C
- Groundwater extraction rate = 20 gpm

Solution:

(a) Use Equation (6.27) to adjust the rate constant:

$$k_2 = k_1 \left(\frac{D_1}{D_2}\right)^{0.3} = (0.9)\left(\frac{2}{3}\right)^{0.3} = 0.80 (gpm)^{0.5}/ft$$

(b) Use Equation (6.25) to determine the surface loading rate, Q/A:

$$\frac{C_{out}}{C_{in}} = \exp[-kD(Q/A)^{-0.5}] = \frac{0.1}{4} = \exp[-(0.80)(3)(Q/A)^{-0.5}]$$

$$Q/A = 0.423 \text{ gpm/ft}^2$$

(c) Use Equation (6.26) to determine the required cross-sectional area:

$$A_{bioreactor} = \frac{Q}{\text{surface loading rate}} = \frac{20 \text{ gpm}}{0.423 \text{ gpm/ft}^2} = 47.2 \text{ ft}^2$$

Diameter of the bioreactor packing = $(4 \times A/\pi)^{1/2}$
$= (4 \times 47.2/\pi)^{1/2} = 7.76$ ft
So, $d = 8$ ft

(d) Assuming the packing material only occupies a small fraction of the total reactor volume, the hydraulic retention time can be estimated by:

hydraulic residence time $= (V/Q) = (Ah)/Q$

$= (47.2 \text{ ft}^2)(3 \text{ ft}) \div [20 \text{ gpm}/(1 \text{ ft}^3/7.48 \text{ gal})] = 53 \text{ min}$

Discussion:

It is relatively difficult for the effluent of a bioreactor to meet the discharge limit in the ppb level. Activated-carbon adsorbers may be needed as a polisher to treat the effluent of the bioreactor before discharge.

6.6 *In Situ* Groundwater Remediation

6.6.1 Description of the *In Situ* Bioremediation Process

Biological *in situ* treatment of organic COCs in aquifers is usually accomplished by enhancing activities of indigenous subsurface microorganisms. Most of the *in situ* bioremediation is practiced in the aerobic mode. The microbial activities are enhanced by addition of inorganic nutrients and oxygen into the groundwater plume. The typical process consists of withdrawal of groundwater, addition of oxygen and nutrients, and reinjection of the enriched groundwater through injection wells or infiltration galleries. In addition to extraction of groundwater, addition of oxygen-releasing compounds (ORCs) to the plume can also be used instead.

6.6.2 Addition of Oxygen to Enhance Biodegradation

Groundwater naturally contains low levels of dissolved oxygen (DO). Even if it is fully saturated with air, the saturated dissolved oxygen (DO_{sat}) concentration in groundwater would only be in the neighborhood of 9 mg/L at 20°C. Biodegradation of organic COCs in the plume may need much more oxygen than those dissolved-oxygen molecules present in groundwater.

Addition of oxygen to the groundwater can be done by air sparging or pure oxygen sparging. The oxygen in the injected air can raise the DO to its saturation level of 8 to 10 mg/L. With pure oxygen injection, the DO concentrations of up to 40 to 50 mg/L can be achieved. DO level in the water can also be raised by addition of chemicals such as hydrogen peroxide and ozone. Each mole of hydrogen peroxide in water can dissociate into 0.5 mole of oxygen and 1 mole of water, while 1 mole of ozone in water can dissociate into 1.5 moles of oxygen as:

$$H_2O_2 \rightarrow H_2O + 0.5\ O_2 \tag{6.28}$$

$$O_3 \rightarrow 1.5\ O_2 \tag{6.29}$$

Ozone is 10 times more soluble in water than pure oxygen. Hydrogen peroxide and ozone can be added to the extracted groundwater before it is put back to the impacted aquifer. It should be noted that hydrogen peroxide and ozone are also strong oxidants. In addition to providing oxygen for biodegradation, they can also generate radicals to oxidize COCs and other inorganic and organic compounds present in the aquifer. However, at higher concentration levels, they may become toxic to indigenous aerobic microorganisms and suppress their biological activities [3].

Various enhanced *in situ* biodegradation approaches rely on oxygen-releasing compounds. The common ORCs include calcium and magnesium peroxides that are introduced to the saturated zone in solid or slurry phases. These peroxides release the oxygen to the aquifer when hydrated by groundwater. Magnesium peroxide has been more commonly used than calcium peroxide due to its lower solubility and prolonged release of oxygen. Oxygen amounting to ≈10% of the mass of magnesium peroxide placed in the saturated zone is released to the aquifer over the active period [3].

Example 6.14: Determine the Necessity of Oxygen Addition for *In Situ* Groundwater Bioremediation

A subsurface is impacted by gasoline. The average dissolved-gasoline concentration of the groundwater samples is 20 mg/L. *In situ* bioremediation is being considered for aquifer restoration. The aquifer has the following characteristics:

- Porosity = 0.35
- Organic content = 0.02
- Subsurface temperature = 20°C
- Dry bulk density of aquifer materials = 1.6 g/cm^3
- DO concentration in the aquifer = 4.0 mg/L

Illustrate that the addition of oxygen to the aquifer is necessary to support biodegradation of the intruded gasoline.

Strategy:

The gasoline compounds in the saturated zone will be dissolved in the groundwater or adsorbed onto the surface of the aquifer materials (assuming the free-product phase is absent). Since only the gasoline concentration in the groundwater is known, we have to estimate the amount of gasoline adsorbed on the aquifer materials by using the partition equation discussed in Chapter 2. In addition, gasoline is a mixture of many compounds, and no specific physicochemical data of gasoline are available; toluene, one of the common constituents, will be used to represent gasoline in this example.

Solution:

Basis: 1 m³ of aquifer

(a) From Table 2.5,

Log K_{ow} = 2.73 (toluene)

Use Equation (2.28) to find K_{oc}:

$K_{oc} = 0.63K_{ow} = 0.63 (10^{2.73}) = (0.63)(537) = 338$

Use Equation (2.26) to find K_p:

$K_p = f_{oc}K_{oc} = (0.02)(338) = 6.8$ L/kg

Use Equation (2.25) to find toluene concentration adsorbed onto the solid:

$S = K_pC = (6.8$ L/kg$)(20$ mg/L$) = 136$ mg/kg

(b) Determine the total mass of gasoline present in the aquifer (per m³):

Mass of the aquifer materials = $(1$ m³$)(1,600$ kg/m³$) = 1,600$ kg

Mass of gasoline adsorbed on the solid surface = $(S)(M_s)$

= $(136$ mg/kg$)(1,600$ kg$) = 217,600$ mg = 218 g

Pore volume of the aquifer = $V \times \phi$

= $(1$ m³$)(35\%) = 0.35$ m³ = 350 L

Mass of gasoline dissolved in the groundwater = $(C)(V_l)$

= $(20$ mg/L$)(350$ L$) = 7,000$ mg = 7.0 g

Total mass of gasoline in the aquifer = dissolved + adsorbed

= $7.0 + 218 = 225$ g of gasoline/m³ of aquifer

(c) The amount of oxygen present in the groundwater = $(V_l)(DO)$

= $(350$ L$)(4$ mg/L$) = 1,080$ mg = 1.08 g

(d) Use the 3.08 ratio to determine the oxygen requirements for complete oxidation of gasoline present (see Section 5.4.4 for details):

Oxygen requirement = $(3.08)(225) = 693 \text{ g} \gg 1.08 \text{ g}$

As demonstrated, the oxygen contained in the groundwater of the aquifer is negligible, when compared to the amount of oxygen required for complete aerobic biodegradation of the gasoline present.

(e) If the groundwater is brought to the surface and aerated with air, the saturated dissolved-oxygen concentration in water at 20°C is approximately 9 mg/L. When this groundwater is recharged back to the impacted zone, the maximum amount of additional oxygen added to the groundwater per m^3 of the aquifer can be found as:

The amount of oxygen added by water saturated with air = $(V_l)(DO_{sat})$

$= (350 \text{ L})(9 \text{ mg/L}) = 3{,}150 \text{ mg} = 3.15 \text{ g}$

Amount of oxygen-enriched water needed to meet the oxygen demand (expressed as the number of pore volumes of the plume) = $(693/3.15) = 247$

Discussion:

1. As shown in part (e), the plume has to be flushed at least 247 times with air-saturated water to meet the oxygen requirement.

2. If the extracted water is aerated with pure oxygen, the saturated DO will be approximately five times higher and the required flushing will be five times less.

3. Fraction of gasoline in the dissolved phase

= (mass of gasoline dissolved)/(total gasoline in the aquifer)

= $(7)/(225) = 3.1\%$

This shows that gasoline dissolved in the groundwater only accounts for a small portion of its total mass in the aquifer.

Example 6.15: Determine the Effectiveness of Hydrogen Peroxide Addition as an Oxygen Source for Bioremediation

As illustrated in Example 6.14, it would take a tremendous amount of water, whether it is saturated with air or pure oxygen, to meet the oxygen demand for *in situ* groundwater bioremediation. Addition of hydrogen peroxide becomes a popular alternative. Because of the biocidal potential of hydrogen peroxide, the maximum hydrogen peroxide in the injected water is often kept below 1,000 mg/L for *in situ* bioremediation applications. Determine the amount of oxygen that 1,000 mg/L of hydrogen peroxide can provide.

Solution:

(a) From Equation (6.28), 1 mole of hydrogen peroxide can yield a half mole of oxygen:

$$H_2O_2 \rightarrow H_2O + 0.5\, O_2$$

Molecular weight of hydrogen peroxide (H_2O_2)

$$= (1 \times 2) + (16 \times 2) = 34$$

Molecular weight of oxygen (O_2) $= 16 \times 2 = 32$

(b) Molar concentration of 1,000 mg/L hydrogen peroxide

$$= (1{,}000 \text{ mg/L}) \div (34{,}000 \text{ mg/mole}) = 29.4 \times 10^{-3} \text{ mole/L}$$

Molar concentration of oxygen (assume 100% dissociation of hydrogen peroxide)

$$= 29.4 \times 10^{-3} \text{ mole } H_2O_2/L \times (0.5 \text{ mole } O_2/\text{mole } H_2O_2)$$
$$= 14.7 \times 10^{-3} \text{ mole/L}$$

Mass concentration of oxygen in water from hydrogen peroxide addition

$$= (14.7 \times 10^{-3} \text{ mole/L}) \times 32{,}000 \text{ mg/mole} = 470 \text{ mg/L}$$

6.6.3 Addition of Nutrients to Enhance Biodegradation

Nutrients for microbial activity usually exist in the subsurface. However, with the presence of organic COCs, additional nutrients are often needed to support the bioremediation. The nutrients to enhance microbial growth are assessed primarily on the nitrogen and phosphorus requirements of the microorganisms. The suggested C:N:P molar ratio is 120:10:1, as shown in Table 5.2. The nutrients are typically added at concentrations ranging from 0.005% to 0.02% by weight [4].

Example 6.16: Determine the Nutrient Requirement for *In Situ* Groundwater Bioremediation

A subsurface is impacted by gasoline. The average dissolved-gasoline concentration of the groundwater samples is 20 mg/L. *In situ* bioremediation is being considered for aquifer restoration. The aquifer has the following characteristics:

- Porosity $= 0.35$
- Organic content $= 0.02$
- Subsurface temperature $= 20°C$
- Dry bulk density of aquifer materials $= 1.6 \text{ g/cm}^3$

Assuming no nutrients are available in the groundwater for bioremediation and the optimal molar C:N:P ratio has been determined to be 100:10:1, determine the amount of nutrients needed to support the biodegradation of gasoline. If the plume is to be flushed with 100 pore volumes of oxygen- and nutrient-enriched water, what would be the required nutrient concentration in the water for reinjection?

Solution:

Basis: 1 m^3 of aquifer

(a) From Example 6.14, the total mass of gasoline in the aquifer = 225 g/m^3

(b) Assume that the gasoline has a formula the same as heptane, C_7H_{16}:

MW of gasoline = $7 \times 12 + 1 \times 16 = 100$

Moles of gasoline = $225/100 = 2.25$ g-mole

(c) Determine the number of moles of C:

Since there are 7 carbon atoms in each gasoline molecule, as indicated by its formula, C_7H_{16}, then:

Moles of C = $(2.25)(7) = 15.8$ g-mole

(d) Determine the number of moles of N needed (using the C:N:P ratio of 100:10:1):

Moles of N needed = $(10/100)(15.8) = 1.58$ g-mole

Amount of nitrogen needed = $1.58 \times 14 = 22.1$ g/m^3 of aquifer

Moles of $(NH_4)_2SO_4$ needed = $1.58 \div 2$

$= 0.79$ g-mole (each mole of ammonium sulfate contains two moles of N)

Amount of $(NH_4)_2SO_4$ needed = $(0.79)[(14+4)(2) + 32 + (16)(4)]$

$= 104$ g/m^3 of aquifer

(e) Determine the number of moles of P needed (using the C:N:P ratio of 100:10:1):

Moles of P needed = $(1/100)(15.8) = 0.158$ g-mole

Moles of $Na_3PO_4 \cdot 12H_2O$ needed = 0.158 g-mole

Amount of phosphorus needed = $0.158 \times 31 = 4.9$ g/m^3 of aquifer

Amount of $Na_3PO_4 \cdot 12H_2O$ needed

$= (0.158)[(23)(3) + 31 + (16)(4) + (12)(18)]$

$= 60$ g/m^3 of aquifer

(f)   The total nutrient requirement = 104 + 60 = 164 g/m³ of aquifer

Void space of the aquifer = $V \times \phi$

$$= (1 \text{ m}^3)(35\%) = 0.35 \text{ m}^3 = 350 \text{ L}$$

Total volume of water that is equivalent to 100 pore volumes = (100)(350) = 35,000 L

The minimum required nutrient concentration = 164 g ÷ 35,000 L

$$= 4.7 \times 10^{-3} \text{ g/L} \approx 0.0005\% \text{ by wt.}$$

Discussion:

The concentration, 0.0005% by wt., is the theoretical amount. In real applications, one may want to add more to compensate the loss due to adsorption to the aquifer material before reaching the plume. This will make the nutrient concentration fall in the typical range of 0.005% to 0.02% by weight.

6.7 Air Sparging

6.7.1 Description of the Air-Sparging Process

Air sparging is an *in situ* remediation technology that involves the injection of air (sometimes pure oxygen) into the saturated zone. The injected air travels through the aquifer, moves upward through the capillary fringe and the vadose zone, and is then collected by the vadose-zone soil-venting network. The injected air (or oxygen) serves the following functions: (1) volatilizes the dissolved VOCs in the groundwater, (2) supplies oxygen to the aquifer for bioremediation, (3) volatilizes the VOCs in the capillary zone as it moves upward, and (4) volatilizes the VOCs in the vadose zone.

6.7.2 Oxygen Addition from Air Sparging

As illustrated in the previous sections, the amount of oxygen carried back to the aquifer by the water, which has been saturated with air or pure oxygen, cannot meet the oxygen demand for *in situ* bioremediation. An air-sparging process continuously brings air (or oxygen) directly into the plume. Consequently, supplying oxygen to the plume to support aerobic biodegradation is one of the main functions of air sparging. Oxygen-transfer efficiency (E) is often used to evaluate the efficacy of aeration, and it is defined as:

$$\text{oxygen-transfer efficiency } (E) = \frac{\text{rate of oxygen dissolution}}{\text{rate of oxygen applied}} \tag{6.30}$$

Many studies have been conducted on oxygen-transfer efficiencies in aeration of water and wastewater treatment, but little information is available with regard to air sparging of aquifers impacted by organics. The oxygen-transfer efficiencies should depend on factors such as injection pressure, the depth of the injection point in the aquifer, and characteristics of the geological formation, to name a few.

Example 6.17: Determine the Rate of Oxygen Addition by Air Sparging

Three air-sparging wells were installed into the plume of an aquifer impacted by hydrocarbons. The injection air flow rate into each well is 5 ft³/min. Assuming the oxygen-transfer efficiency is 10%, determine the rate of oxygen addition to the aquifer through each sparging well. What would be the equivalent injection rate of water that is saturated with air?

Solution:

(a) The oxygen concentration in the ambient air is approximately 21% by volume, which is equal to 210,000 ppmV. Equations (2.1) or (2.2) can be used to convert it to a mass concentration:

$$1\ \text{ppmV} = \frac{\text{MW}}{385} \times 10^{-6} \quad [\text{lb/ft}^3] \quad \text{at } 68\,^\circ\text{F}$$

$$= \frac{32}{385} \times 10^{-6} = 0.083 \times 10^{-6}\ \text{lb/ft}^3 \qquad (2.2)$$

Therefore, $210,000\ \text{ppmV} = (210,000)(0.083 \times 10^{-6}) = 0.0175\ \text{lb/ft}^3$

(b) The rate of oxygen injected in each well = $(G)(Q)$

$= (0.0175\ \text{lb/ft}^3)(5\ \text{ft}^3/\text{min}) = 0.0875\ \text{lb/min} = 126\ \text{lb/day}$

The rate of oxygen dissolved into the plume through air injection in each well (using Equation 6.30) = $(126\ \text{lb/day})(10\%) = 12.6$ lb/day

(c) The dissolved-oxygen concentration of the air-saturated reinjection water is approximately 9 mg/L at 20°C. The required water-reinjection rate to supply 1.26 lb/day of oxygen can be found as:

$12.6\ \text{lb/day} = (12.6\ \text{lb/day})(454,000\ \text{mg/lb}) = (Q)(C) = (Q)(9\ \text{mg/L})$
Thus, $Q = 635,600\ \text{L/day} = 116.6\ \text{gpm}$

Discussion:

1. The oxygen-transfer efficiency of 10% means that only 10% of the total oxygen sparged into the aquifer will be dissolved into the aquifer. But, 90% of the oxygen injected can serve as the oxygen source for bioremediation in the vadose zone.

2. Despite the relatively low oxygen-transfer efficiency in this example, the air sparging still adds a significant amount of oxygen
   to the aquifer. With regard to oxygen addition, an air injection
   rate of 5 ft³/min with an oxygen-transfer efficiency of 10% is
   equivalent to reinjection of air-saturated water at 116.6 gpm.

6.7.3 Injection Pressure of Air Sparging

Injection pressure is an important component for design of an air-sparging
system. The applied air injection pressure should overcome at least (1) the
hydrostatic pressure corresponding to the water-column height above the
injection point and (2) the air-entry pressure, which is equivalent to the capillary pressure necessary to induce air into the saturated media.

$$P_{injection} = P_{hydrostatic} + P_{capillary} \qquad (6.31)$$

Reported values of injection pressures range from 1 to 8 psig [5].
  The following procedure can be used to determine the minimum air injection pressure:

Step 1: Determine the water-column height above the injection point.
  Convert the water-column height to pressure units by using the
  following formula:

$$P_{hydrostatic} = \rho g \, h_{hydrostatic} \qquad (6.32)$$

where ρ is the mass density of water and g is the gravitational
constant (32.2 ft/s²).

Step 2: Use Table 2.2 to estimate pore radius of the aquifer media and
  then use Equation (2.11) to determine height of capillary rise (or
  obtain the capillary height from Table 2.2 directly). Convert the
  capillary height to the capillary pressure by using the following formula:

$$P_{capillary} = \rho g \, h_{capillary} \qquad (6.33)$$

Step 3: The minimum air injection pressure is the sum of these two pressure components ($P_{hydrostatic} + P_{capillary}$) as stated in Equation (6.31).

Information needed for this calculation:

- Depth of the injection point, $h_{hydrostatic}$
- Mass density of water, ρ
- Geology of the aquifer material or the pore size of the matrix

Example 6.18: Determine the Required Injection Pressure of Air Sparging

Three air-sparging wells were installed into the plume of the aquifer described in Example 6.17. The injection air flow rate into each well is 5 ft³/ min. The height of the water column above the air injection point is 10 ft. The aquifer matrix consists mainly of coarse sand. Determine the minimum air injection pressure required. Also, for the purpose of comparison, determine the air injection pressure if the aquifer formation is clayey.

Solution:

(a) Use Equation (6.33) to convert the water-column height to pressure units as:

$$P_{hydrostatic} = \rho g \, h_{hydrostatic} = \left(62.4 \, \frac{lb_m}{ft^3}\right)\left(32.2 \, \frac{ft}{s^2}\right)(10 \, ft)\left[\frac{lb_f}{32.2 \, lb_m - ft/s^2}\right]$$

$$= 624 \, \frac{lb_f}{ft^2} = 4.33 \, \frac{lb_f}{in^2} = 4.33 \, psi$$

   Note: The density of water at 60°F is 62.4 lb_m/ft³. In other words, the specific weight of water is 62.4 lb_f/ft³. The water-column height of 33.9 ft at 60°F is equivalent to a pressure of 1 atm or 14.7 psi.

(b) From Table 2.2, pore radius of fine-sand media is 0.05 cm. Use Equation (2.11) to determine the height of capillary rise:

$$h_c = \frac{0.153}{r} = \frac{0.153}{0.05} = 3.06 \, cm = 0.1 \, ft$$

   Use the discussion in part (a) to convert the capillary rise to the capillary pressure:

$$P_{capillary} = \left(\frac{0.1 \, ft}{33.9 \, ft}\right)(14.7 \, psi) = 0.04 \, psi$$

(c) Use Equation (6.31) to determine the minimum air injection pressure:

$$P_{injection} = P_{hydrostatic} + P_{capillary} = 4.33 + 0.04 = 4.37 \, psig$$

(d) If the aquifer formation is clayey, then the pore radius is 0.0005 cm (from Table 2.2). Use Equation (2.11) to determine the height of the capillary rise:

$$h_c = \frac{0.153}{r} = \frac{0.153}{0.0005} = 306 \, cm = 10 \, ft$$

Use the discussions in part (a) to convert the capillary rise to the capillary pressure:

$$P_{capillary} = \left(\frac{10\ ft}{33.9\ ft}\right)(14.7\ psi) = 4.33\ psi$$

Use Equation (6.31) to determine the minimum air injection pressure:

$$P_{injection} = P_{hydrostatic} + P_{capillary} = 4.33 + 4.33 = 8.66\ psig$$

Discussion:

1. The actual air injection pressure should be larger than the minimum air injection pressure calculated here to cover the system pressure loss such as head loss in the pipeline, fittings, and injection head.

2. For sandy aquifers, the air entry pressure is negligible compared to the hydrostatic pressure. However, for clayey aquifers, the entry pressure is of the same order of magnitude as the hydrostatic pressure.

3. The calculated injection pressures are in the ballpark of the reported field values, 1–8 psig.

6.7.4 Power Requirement for Air Injection

Theoretical horsepower requirements ($hp_{theoretical}$) of gas compressors for an ideal gas undergoing an isothermal compression (PV = constant) can be expressed as [6]:

$$hp_{theoretical} = 3.03 \times 10^{-5} P_1 Q_1 \ln \frac{P_2}{P_1} \tag{6.34}$$

where
$P_1 =$ intake pressure, lb_f/ft^2
$P_2 =$ final delivery pressure, lb_f/ft^2
$Q_1 =$ air flow rate at the intake condition, ft^3/min
For an ideal gas undergoing an isentropic compression (PV^k = constant), the following equation applies for single-stage compressor [6]:

$$hp_{theoretical} = \frac{3.03 \times 10^{-5} k}{k-1} P_1 Q_1 \left[\left(\frac{P_2}{P_1}\right)^{(k-1)/k} - 1\right] \tag{6.35}$$

where k is the ratio of specific heat of gas at constant pressure to specific heat of gas at constant volume. For air-sparging applications, it is appropriate to use $k = 1.4$.

For reciprocating compressors, the efficiencies (*E*) are generally in the range of 70% to 90% for isentropic and 50% to 70% for isothermal compression. The actual horsepower requirement can be found as:

$$hp_{actual} = \frac{hp}{E} \tag{6.36}$$

Example 6.19: Determine the Power Requirement for Air Sparging

Three air-sparging wells were installed into the contaminant plume of the aquifer described in Example 6.17. The injection air flow rate into each well is 5 ft³/min. A compressor is to serve all three wells. Head loss of the piping system and the injection head was found to be 1 psi. Using the calculated air injection pressure from Example 6.18, determine the required horsepower of the compressor.

Solution:

(a) The required injection pressure = the final delivery of the compressor, P_2
    = minimum injection pressure + head loss = 4.37 + 1.0
    = 5.37 psig
    = (5.37 + 14.7) psi = 20.1 psi = (20.1)(144) = 2,890 lb/ft²

(b) Assuming isothermal expansion, use Equation (6.34) to determine the theoretical power requirement as:

$$hp_{theoretical} = 3.03 \times 10^{-5} P_1 Q_1 \ln \frac{P_2}{P_1}$$

$$= 3.03 \times 10^{-5} [(14.7)(144)][(3)(5)] \ln \frac{2,890}{(14.7)(144)} = 0.3 \text{ hp}$$

Assuming an isothermal efficiency of 60%, the actual horsepower required is determined by using Equation (6.36):

$$hp_{actual} = \frac{hp_{theoretical}}{E} = \frac{0.3}{50\%} = 0.5 \text{ hp}$$

(c) Assuming isothermal expansion, use Equation (6.35) to determine the theoretical power requirement as:

$$hp_{theoretical} = \frac{3.03 \times 10^{-5} k}{k-1} P_1 Q_1 \left[\left(\frac{P_2}{P_1} \right)^{(k-1)/k} - 1 \right]$$

$$= \frac{(3.03 \times 10^{-5})(1.4)}{1.4-1} [(14.7)(144)][(5.0)(3)] \left[\left(\frac{2,890}{(14.7)(144)} \right)^{(1.4-1)/1.4} - 1 \right]$$

$$= 0.31 \text{ hp}$$

Assuming an isentropic efficiency of 80%, the actual horsepower required is determined by using Equation (6.36) as:

$$hp_{actual} = \frac{hp_{theoretical}}{E} = \frac{0.31}{80\%} = 0.4 \text{ hp}$$

Discussion:

The energy necessary for an isentropic compression is generally greater than that for an equivalent isothermal compression. However, the difference between the inlet and final discharge pressures in most air-sparging applications is relatively small. Consequently, the theoretical power requirements for the isothermal and isentropic compressions should be very similar, as illustrated in this example.

6.8 Biosparging

Biosparging is an *in situ* remediation technology for aquifers impacted by biodegradable organics. It enhances the biological activities of indigenous microorganisms to biodegrade organic constituents in the saturated zone by injecting air (or pure oxygen) and nutrients, if needed, into the plume. In addition, to reduce concentrations of COCs in the aquifer, biosparging may also reduce those in the capillary zone.

The biosparging process is similar to the air-sparging process. However, the COC removal mechanism of air sparging is mainly volatilization, while that of biosparging is enhancement of *in situ* biodegradation. In general, some degree of volatilization and biodegradation occurs when either air sparging or biosparging is used. Biosparging can be more effective on semi-volatile compounds than air sparging. The air injection rate for biosparging is typically smaller than that of air sparging, and the air can be injected on an intermittent basis, just to support the biological activities. However, when volatile constituents are present, biosparging often needs to be combined with soil-vapor extraction or bioventing for concern of fugitive emission.

Design calculations for biosparging are essentially the same as those for air sparging. Please refer to the discussion and examples presented in Section 6.7.

6.9 Metal Removal by Chemical Precipitation

Elevated heavy metal concentrations may occur in extracted groundwater or in wastewater streams. Chemical precipitation is one of the common methods to remove inorganic heavy metals from groundwater or wastewater. The hydroxides of heavy metals are formed at high pH and are usually insoluble. Lime or caustic soda addition is often made to precipitate the metals. The solubility of metal hydroxides is sensitive to pH, and the reaction can be expressed in a general form:

$$M(OH)_n \leftrightarrow M^{n+} + nOH^- \tag{6.37}$$

where M represents the heavy metal, OH^- is the hydroxide ion, and n is the valence of the metal.

The equilibrium equation can be written as:

$$K_{sp} = [M^{n+}][OH^-]^n \tag{6.38}$$

where K_{sp} is the equilibrium constant (often called the *solubility product*), $[M^{n+}]$ is the molar concentration of the heavy metal, and $[OH^-]$ is the molar concentration of hydroxide ions. For example, the K_{sp} values for $Cr(OH)_3$, $Fe(OH)_3$, and $Mg(OH)_2$ at 25°C are 6×10^{-31} M^4, 6×10^{-36} M^4, and 9×10^{-12} M^3, respectively.

Example 6.20: Chemical Precipitation for Magnesium Removal

Sodium hydroxide is added to a continuous-flow stirred tank reactor (CFSTR) to remove magnesium ion from an extracted groundwater stream ($Q = 150$ gpm). The temperature of the reactor is kept at 25°C and pH = 11. The influent Mg^{2+} concentration is 100 mg/L. If the solids are settled to 10% by weight, estimate

(a) Mg^{2+} concentration in the treated effluent (mg/L)
(b) rate of $Mg(OH)_2$ produced (lb/day)
(c) rate of sludge produced (lb sludge/day)

(Note: The solubility product of $Mg(OH)_2$ is 9×10^{-12} M^3 at 25°C; MW of Mg = 24.3.)

Solution:
(a) Write the reaction of precipitation first:

$$Mg(OH)_2 \leftrightarrow Mg^{2+} + 2OH^-$$

At pH = 11, the hydroxide concentration $[OH^-]$ is equal to 10^{-3}. Use the solubility product equation to determine the magnesium concentration as:

$$K_{sp} = [Mg^{2+}][OH^-]^2 = 9 \times 10^{-12} = [Mg^{2+}][10^{-3}]^2$$

$$[Mg^{2+}] = 9 \times 10^{-6} M = (9 \times 10^{-6} \text{ mole/L})(24.3 \text{ g/mole})$$

$$= 2.19 \times 10^{-4} \text{ g/L} = 0.22 \text{ mg/L}$$

(b)   As shown in part (a), one mole of $Mg(OH)_2$ is formed for each mole of Mg^{2+} removed. Since the molecular weight of $Mg(OH)_2$ is equal to 58.3, the rate of $Mg(OH)_2$ produced can be found as:

Rate of $Mg(OH)_2$ produced = (Rate of Mg^{2+} removed)(58.3/24.3)

$= \{[Mg^{2+}]_{in} - [Mg^{2+}]_{out}\} \times (Q) \times (58.3/24.3)$

$= [(100 - 0.22) \text{ mg/L}] \times [(150 \text{ gpm})(3.785 \text{ L/gal})] \times (58.3/24.3)$

$= 136,000 \text{ mg/min} = 136 \text{ g/min} = 431 \text{ lb/day}$

(c)   Since the solids are settled to 1% by weight, the rate of sludge production can be found as:

Rate of sludge produced = Rate of $Mg(OH)_2$ produced ÷ 10%

$= 431 \text{ lb/day} \div 10\% = 4,310 \text{ lb/day}$

6.10 *In Situ* Chemical Oxidation

In situ chemical oxidation (ISCO) involves the introduction of a chemical oxidant into the subsurface to transform COCs in soil or groundwater into less harmful compounds. ISCO is predominantly used to address COCs in the source area so that the mass flux to the groundwater plume can be reduced. Consequently, it can shorten anticipated cleanup times for natural attenuation and other remedial options [3].

The approaches of using ISCO for impacted aquifers are essentially the same as those for the vadose zone. For the background information on types of oxidants as well as oxidant demands, please refer to Section 5.6 for details. The main difference between groundwater remediation versus vadose zone remediation is that the applied oxidant needs to be delivered to the saturation zone, in which all the pore space is filled with water. This section illustrates one example that is related to application of ISCO to the saturated zone.

Example 6.21: Determine the Stoichiometric Amount of Oxidant

The underlying aquifer at a site is impacted by perchloroethylene (PCE). The source area for the groundwater plume has been determined to be a zone with an area of 20 m² and a thickness of 2 m within the aquifer. The average PCE concentration of groundwater samples taken from this zone is 400 mg/L.

The aquifer has the following characteristics:

- Porosity = 0.35
- Organic content = 0.02
- Subsurface temperature = 20°C
- Dry bulk density of aquifer materials = 1.6 g/cm³

In situ chemical oxidation is considered as one of the remedial alternatives. Determine the stoichiometric amount of potassium permanganate that needs to be delivered to the zone. What would be the amount if sodium persulfate is used?

Solution:

Basis: 1 m³ of aquifer

(a) Determine log K_{ow} from Table 2.5:

$$\log K_{ow} = 2.6$$

Use Equation (2.28) to find K_{oc}:

$$K_{oc} = 0.63K_{ow} = 0.63(10^{2.6}) = (0.63)(398) = 251$$

Use Equation (2.26) to find K_p:

$$K_p = f_{oc}K_{oc} = (0.02)(251) = 5.02 \text{ L/kg}$$

Use Equation (2.25) to find PCE concentration adsorbed onto the solid:

$$S = K_pC = (5.02 \text{ L/kg})(400 \text{ mg/L}) = 2,006 \text{ mg/kg}$$

(b) Determine the total mass of PCE present in the aquifer (per m³):

Mass of the aquifer materials = (1 m³)(1,600 kg/m³) = 1,600 kg

Mass of PCE adsorbed on the solid surface = $(S)(M_s)$

= (2,006 mg/kg)(1,600 kg) = 3,210,000 mg = 3,210 g

Pore volume of the aquifer = $V \times \phi$

= (1 m³)(35%) = 0.35 m³ = 350 L

Mass of PCE dissolved in the groundwater = $(C)(V_l)$

= (400 mg/L)(350 L) = 140,000 mg = 140 g

Total mass of PCE in the aquifer = dissolved + adsorbed

= 140 + 3,210 = 3,350 g of PCE/m³ of aquifer

(c) MW of PCE (C_2Cl_4) = (12)(2) + (35.5)(4) = 166

MW of potassium permanganate ($KMnO_4$) = (39)(1) + (55)(1) + (16)(4) = 158

Moles of PCE = 3,350 g/m³ of aquifer

= (3,350 g ÷ 166 g/mole)/m³ of aquifer

= 20.2 moles of PCE/m³ of aquifer

As shown in Equation (5.42), the stoichiometric requirement to oxidize PCE is 4/3 mole permanganate per mole of PCE.

Stoichiometric amount of $KMnO_4$ needed per m³ of aquifer

= (4 moles of $KMnO_4$/3 moles of PCE) × (20.2 moles of PCE/ m³ of aquifer)

$= 26.9$ moles of $KMnO_4/m^3$ of aquifer

$= (26.9$ mole $\times 158$ g/mole $KMnO_4)/m^3$ of aquifer

$= 4{,}250$ g $KMnO_4/m^3$ of aquifer $= 4.25$ kg $KMnO_4/m^3$ of aquifer

Stoichiometric amount of $KMnO_4$ needed for the entire zone $(40\ m^3)$

$= (4.25$ kg $KMnO_4/m^3$ of aquifer$) \times 40\ m^3$

$= 170$ kg $KMnO_4$

(d) MW of sodium persulfate $(Na_2S_2O_8) = (23)(2) + (32)(2) + (16)(8) = 238$

As shown in Table 5.3 and as discussed, the stoichiometric requirement of sodium persulfate will be 1.5 times of that of potassium permanganate.

Stoichiometric amount of $Na_2S_2O_8$ (per m^3 of aquifer)

$= (3$ moles of $Na_2S_2O_8/2$ moles of $KMnO_4) \times (26.9$ moles of $KMnO_4/m^3$ of aquifer$)$

$= (40.35$ moles $Na_2S_2O_8/m^3$ of aquifer$)$

$= (40.35$ moles $\times 238$ g/mole $Na_2S_2O_8/m^3$ of aquifer$)$

$= 9{,}600$ g $Na_2S_2O_8/m^3$ of aquifer $= 9.6$ kg $Na_2S_2O_8/m^3$ of aquifer

Stoichiometric amount of $Na_2S_2O_8$ needed for the entire zone $(40\ m^3)$

$= (9.6$ kg $Na_2S_2O_8/m^3$ of aquifer$) \times 40\ m^3$

$= 384$ kg $Na_2S_2O_8$

6.11 Advanced Oxidation Process

Advanced oxidation process (AOP) refers to an oxidation process assisted by ultraviolet (UV) irradiation. In AOP, high-power lamps emit UV radiation through quartz sleeves into impacted groundwater. An oxidizing agent, typically hydrogen peroxide, ozone, or a combination of these two is added. The oxidizing agent is activated by the UV light to form hydroxyl radicals, which have a very strong oxidizing power. These radicals destroy the organic COCs in the impacted groundwater.

In a typical AOP, oxidizing reagents are often injected and mixed using metering pumps and in-line static mixers. The groundwater then flows sequentially through one or more UV reactors. The reactors are often considered as plug-flow type, and the reactions follow first-order kinetics. Equation (4.24) describes the relationship among the influent concentration, effluent concentration, retention time, and reaction rate constant for plug-flow reactors. It is repeated here for the AOP reactors as:

$$\frac{C_{out}}{C_{in}} = e^{-k(V/Q)} = e^{-k\tau} \tag{6.39}$$

where C is the COC concentration in groundwater, V is the reactor volume, Q is the groundwater flow rate, k is the rate constant, and τ is the hydraulic retention time.

Another design approach is to use electrical energy per order of destruction (EE/O) to scale up an AOP reactor. An EE/O of 5 kWh/1,000 gal/order of COC destruction means that it will take 5 kWh of energy to reduce the COC concentration from 1 ppm to 0.1 ppm in 1,000 gal of groundwater. It will take another 5 kWh of energy to reduce the concentration from 0.1 ppm to 0.01 ppm. It should be noted that the value of EE/O is specific to the groundwater and COCs treated.

Example 6.22: Sizing the Reactor for an Advanced Oxidation Process

UV/ozone treatment is selected to remove trichloroethylene (TCE) from an extracted groundwater stream (TCE concentration = 400 ppb). A pilot study was conducted and found that, with a hydraulic retention of 2 min, the system could reduce TCE concentration from 400 ppb to 16 ppb. However, the discharge limit for TCE is 3.2 ppb. Assuming the reactors are of ideal plug-flow type and the reaction is first-order, how many reactors would you recommend to use?

Solution:

(a) Use Equation (6.39) to determine the reaction rate constant:

$$\frac{C_{out}}{C_{in}} = \exp[-k(\tau)] = \frac{16}{400} = \exp[-2k]$$

So, $k = 1.61/\text{min}$

(b) Use Equation (6.39) to determine the required retention time to reduce the TCE concentration below the discharge limit:

$$\frac{C_{out}}{C_{in}} = \exp[-k(\tau)] = \frac{3.2}{400} = \exp[-1.61(\tau)]$$

$\tau = 3.0$ min. Thus, it requires two reactors.

(c) Use Equation (6.39) to determine the final effluent TCE concentration ($\tau = 4$ min because two reactors were used):

$$\frac{C_{out}}{C_{in}} = \exp[-k(\tau)] = \frac{C_{out}}{400} = \exp[-1.61(4)]$$

$C_{out} = 0.64$ ppb

Discussion:

1. For PFRs, the final concentration from two identical reactors in series is the same as that from two identical reactors in parallel.

2. A pilot-scale test to determine the removal efficiency and the reaction rate constant is always recommended for AOPs.

Example 6.23: Sizing the Reactor for an Advanced Oxidation Process

UV/ozone treatment is selected to remove TCE from an extracted ground-water stream ($Q = 100$ gpm, TCE concentration $= 400$ ppb). A pilot study was conducted and found the electrical energy per order to be 6 kWh/1,000 gal/ log of TCE reduction for this type of groundwater. What would be the daily energy requirement to reduce TCE concentration from 400 ppb to 16 ppb?

Solution:

(a) The reduction from 400 to 16 is equal to $\log(400/16) = 1.4$ logs

(b) The total volume of water treated per day
   $= (100 \text{ gal/min}) \times (1{,}440 \text{ min/day}) = 144{,}000$ gal

(c) Daily energy required $= (1.4 \text{ logs}) \times (6 \text{ kWh/1,000 gal/log reduction}) \times (144{,}000 \text{ gal})$
   $= 1{,}210$ kWh

Discussion:

If the cost of electricity is $0.15/kWh, the energy cost will be $181.5/day.

References

1. Javandel, I., and Chin-Fu Tsang. 1986. Capture-zone type curves: A tool for aquifer cleanup. *Groundwater* 24 (5): 616–25.
2. Metcalf & Eddy, Inc. 1991. *Wastewater engineering*. 3rd ed. New York: McGraw-Hill.
3. USEPA. 2004. How to evaluate alternative cleanup technologies for underground storage sites. EPA/510/R-04/002. Washington, DC: Office of Solid Waste and Emergency Response, US EPA.
4. USEPA. 1991. Site characterization for subsurface remediation. EPA/625/R-91/026. Washington, DC: Office of Research and Development, US EPA.
5. Johnson, R.L., P.C. Johnson, D.B. McWhorter, R.E. Hinchee, and I. Goodman. 1993. An overview of *in situ* air sparging. *Ground Water Monitoring Review*, Fall:127–35.
6. Peters, M.S., and K.D. Timmerhaus. 1991. *Plant design and economics for chemical engineers*. 4th ed. New York: McGraw-Hill.

7

VOC-Laden Air Treatment

7.1 Introduction

Remediation of impacted soil/groundwater often results in transfer of organic compounds of concern (COCs) from soil/groundwater into air. The air stream, containing organic COCs, usually needs to be treated before being released to the atmosphere. Development and implementation of an air emission control strategy should be an integral part of the overall remediation program. Air emission control can be expensive, and it may affect the cost-effectiveness of a specific remedial alternative.

Common sources of volatile organic compounds (VOC)-laden off-gas from soil/groundwater remediation activities include soil vapor extraction (SVE), low-temperature thermal desorption, soil washing, solidification/stabilization, air sparging, biosparging, air stripping, and bioremediation. This chapter covers some design calculations for commonly used off-gas treatment technologies, including activated-carbon adsorption, direct incineration, catalytic incineration, internal combustion (IC) engines, and biofiltration.

Much background information in this chapter is taken from three technical articles published by the US Environmental Protection Agency (USEPA) [1–3].

7.2 Activated-Carbon Adsorption

7.2.1 Description of the Activated-Carbon Adsorption Process

Activated-carbon adsorption is one of the most commonly used air pollution control processes for VOC emissions. The process is very effective in removing a wide range of VOCs. The most common form of vapor-phase activated carbon is granular activated carbon (GAC).

Activated carbon has a fixed capacity or a limited number of active adsorption sites. Once the adsorbed COCs occupy most of the available sites, the removal efficiency will drop significantly. If the operation is continued beyond this point, the breakthrough point will be reached and the effluent concentration will rise sharply. Eventually, the activated carbon in service would be saturated, exhausted, or spent when most of the sites are occupied. The spent carbon needs to be regenerated or disposed of.

Two pretreatment processes of the influent air are often required to optimize the performance of GAC systems. The first is cooling, and the other is dehumidification. Adsorption of VOCs is generally exothermic, which is favored by lower temperatures. As a rule of thumb, the waste air stream needs to be cooled down below 130°F. On the other hand, water vapor will compete with VOCs in the waste air stream for available adsorption sites. The relative humidity of the waste air stream generally should be reduced to 50% or less. Taking the off-gas stream from an air stripper as an example, it is typically saturated with water. The air stream may need to be cooled down (e.g., using chiller water) to condense out the moisture and then heated up to some extent (e.g., using an electrical heater) to raise its relative humidity if activated carbon is used to remove the VOCs before discharge.

7.2.2 Sizing Criteria for Granular Activated Carbon

Two common types of vapor-phase activated-carbon systems are (1) canister systems with off-site regeneration and (2) multiple-bed systems with on-site batch regeneration (while some of the adsorption units are in the adsorption cycle and the others are in the regeneration cycle).

Sizing of the GAC systems depends primarily on the following parameters:

- Volumetric flow rate of VOC-laden gas stream
- Concentration or mass loading of VOCs
- Adsorption capacity of the GAC
- Design GAC regeneration frequency

The design flow rate affects the sizing of the cross-sectional area of the GAC unit, the fan and motor, and the duct. The other three parameters (i.e., mass loading, GAC adsorption capacity, and regeneration frequency) affect the size and the number of the units as well as the amount of GAC required for a specific project. Design principles for vapor-phase activated-carbon systems are basically similar to those for liquid-phase activated-carbon systems, as discussed in Section 6.3.

7.2.3 Adsorption Isotherm and Adsorption Capacity

The adsorption capacity of GAC depends on the characteristics of GAC, characteristics of VOCs and their concentration, temperature, and the presence of other species competing for adsorption. At a given temperature, a relationship exists between the mass of the VOC adsorbed per unit mass GAC to the

TABLE 7.1

Empirical Constants for Selected Adsorption Isotherms

Compounds	Adsorption Temperature (°F)	a	m	Range of P_{VOC} (psi)
Benzene	77	0.597	0.176	0.0001–0.05
Toluene	77	0.551	0.110	0.0001–0.05
m-Xylene	77	0.708	0.113	0.0001–0.001
m-Xylene	77	0.527	0.0703	0.001–0.05
Phenol	104	0.855	0.153	0.0001–0.03
Chlorobenzene	77	1.05	0.188	0.0001–0.01
Cyclohexane	100	0.508	0.210	0.0001–0.05
Dichloroethane	77	0.976	0.281	0.0001–0.04
Trichloroethane	77	1.06	0.161	0.0001–0.04
Vinyl chloride	100	0.20	0.477	0.0001–0.05
Acrylonitrile	100	0.935	0.424	0.0001–0.05
Acetone	100	0.412	0.389	0.0001–0.05

Source: [1].

concentration (or partial pressure) of VOC in the waste air stream. For most of the VOCs, the adsorption isotherms can be fitted well by a power curve, also known as the Freundlich isotherms (also see Equation 6.7):

$$q = a(P_{VOC})^m \tag{7.1}$$

where

q = equilibrium adsorption capacity, lb VOC/lb GAC
P_{VOC} = partial pressure of VOC in the waste air stream, psi
a, m = empirical constants

The empirical constants of the Freundlich isotherms for selected VOCs are listed in Table 7.1. It should be noted that the values of these empirical constants are for a specific type of GAC only and should not be used outside the specified range.

The actual adsorption capacity in the field applications should be lower than the equilibrium adsorption capacity. Normally, design engineers take 25% to 50% of the equilibrium value as the design adsorption capacity as a factor of safety. Therefore,

$$q_{design} = (50\%)(q_{theoretical}) \tag{7.2}$$

The maximum amount of COCs that can be removed or held ($M_{removal}$) by a given amount of GAC can be determined as:

$$M_{removal} = (q_{design})(M_{GAC})$$

$$= (q_{design})[(V_{GAC})(\rho_b)] \tag{7.3}$$

where M_{GAC} is the mass, V_{GAC} is the volume, and ρ_b is the bulk density of the GAC, respectively.

The following procedure can be used to determine the adsorption capacity of a GAC adsorber:

Step 1: Determine the theoretical adsorption capacity by using Equation (7.1).

Step 2: Determine the actual adsorption capacity by using Equation (7.2).

Step 3: Determine the amount of activated carbon in the adsorption unit (also called the *adsorber*).

Step 4: Determine the maximum amount of contaminants that can be held by the adsorber by using Equation (7.3).

Information needed for this calculation:

- Adsorption isotherm
- COC concentration in the influent air stream, P_{VOC}
- Volume of the GAC, V_{GAC}
- Bulk density of the GAC, ρ_b

Example 7.1: Determine the Capacity of a GAC Adsorber

The off-gas from a soil-venting project is to be treated by GAC adsorbers. The m-xylene concentration in the off-gas is 800 ppmV. The flow rate out of the vacuum pump is 200 cfm, and the temperature of the air is ambient. Two 1000-lb activated-carbon adsorbers are proposed. Determine the maximum amount of m-xylene that can be held by each GAC adsorber before being exhausted. Use the isotherm data in Table 7.1.

Solution:

(a) Convert the xylene concentration from ppmV to psi as:

$$P_{VOC} = 800 \text{ ppmV} = 800 \times 10^{-6} \text{ atm} = 8.0 \times 10^{-4} \text{ atm}$$
$$= (8.0 \times 10^{-4} \text{ atm})(14.7 \text{ psi/atm}) = 0.0118 \text{ psi}$$

Obtain the empirical constants for the adsorption isotherm from Table 7.1 and then apply Equation (7.1) to determine the equilibrium adsorption capacity as:

$$q = a(P_{VOC})^m = (0.527)(0.0118)^{0.0703} = 0.386 \text{ lb/lb}$$

(b) The actual adsorption capacity can be found by using Equation (7.2) as:

$$q_{design} = (50\%)q_{theoretical} = (50\%)(0.386) = 0.193 \text{ lb/lb}$$

(c) The amount of xylene that can be retained by an adsorber before the GAC becomes exhausted

> = (amount of the GAC)(actual adsorption capacity)
> = (1,000 lb/unit)(0.193 lb xylene/lb GAC) = 193 lb xylene/unit

Discussion:

1. The adsorption capacity of vapor-phase GAC is typically in the neighborhood of 0.1 lb/lb (or 0.1 kg/kg), which is much higher than the adsorption capacity of liquid-phase GAC, typically in the neighborhood of 0.01 lb/lb.

2. Care should be taken to use matching units for P_{voc} and q in the adsorption isotherms.

3. The influent COC concentration in the air stream, not the effluent concentration, should be used in the adsorption isotherms to determine the adsorption capacity.

4. There are two sets of empirical constants for m-xylene; one should always check the applicable range for the empirical constants.

7.2.4 Cross-Sectional Area and Height of GAC Adsorbers

To achieve efficient adsorption, the air flow rate through the activated carbon should be kept as low as possible. The practical design velocity is often ≤60 ft/min, and 100 ft/min is considered as the maximum value. This design parameter can be used to determine the cross-sectional area of the GAC adsorbers (A_{GAC}):

$$A_{GAC} = \frac{Q}{\text{Air flow velocity}} \tag{7.4}$$

where Q is the influent air flow rate. The design height of the adsorber is normally 2 ft or deeper to provide a sufficiently large mass-transfer zone for adsorption.

Example 7.2: Required Cross-Sectional Area of GAC Adsorbers

Referring to the remediation project described in Example 7.1, the 1000-lb GAC units are out of stock. To avoid delay of remediation, off-the-shelf 55-gal activated-carbon units are proposed on an interim basis. The type of GAC in the 55-gal units is the same as that in the 1,000-lb units. The vendor also provided the following information with regard to the units:

- Diameter of carbon packing bed in each 55-gal drum = 1.5 ft
- Height of carbon packing bed in each 55-gal drum = 3 ft
- Bulk density of the activated carbon = 28 lb/ft³

Determine (1) the amount of activated carbon in each 55-gal unit, (2) the amount of xylene that each unit can remove before being exhausted, and (3) the minimum number of the 55-gal units needed.

Solution:

(a) Volume of the activated carbon inside a 55-gal drum = $(\pi r^2)(h)$

$= (\pi)[(1.5/2)^2](3) = 5.3$ ft^3

Amount of the activated carbon inside a 55-gal drum = $(V)(\rho_b)$

$= (5.3 \text{ ft}^3)(28 \text{ lb/ft}^3) = 148$ lb

(b) Amount of xylene that can be retained by a drum before the GAC becomes exhausted

= (amount of the GAC)(actual adsorption capacity)

= (148 lb/drum)(0.193 lb xylene/lb GAC) = 28.6 lb xylene/drum

(c) Assuming a design air flow velocity of 60 ft/min, the required cross-sectional area for the GAC adsorption can be found by using Equation (7.4) as:

$$A_{GAC} = \frac{Q}{\text{Air flow velocity}} = \frac{200}{60} = 3.33 \text{ ft}^2$$

If the adsorption system is tailor-made, then a system with a cross-sectional area of 3.33 ft^2 will do the job. However, the off-the-shelf 55-gal drums are to be used, so we need to determine the number of drums that will provide the required cross-sectional area.

Area of the activated carbon inside a 55-gal drum = (πr^2)

$= (\pi)[(1.5/2)^2] = 1.77$ ft^2/drum

Number of drums in parallel to meet the required hydraulic loading rate

$= (3.33 \text{ ft}^2) \div (1.77 \text{ ft}^2/\text{drum}) = 1.88$ drums

So, use two drums in parallel to provide the required cross-sectional area. The total cross-sectional area of two drums is equal to 3.54 ft^2 ($= 1.77 \times 2$).

Discussion:

1. The bulk density of vapor-phase GAC is typically in the neighborhood of 30 lb/ft^3. The amount of activated carbon in a 55-gal drum is approximately 150 pounds.

2. The minimum number of 55-gal drums for this project is two to meet the air flow velocity requirement. The actual number

of drums should be more to meet the monitoring requirements or the desirable frequency of change-out. If multiple
GAC adsorbers are used, the adsorbers are often arranged
in series and/or in parallel. If two adsorbers are arranged in
series, the monitoring point can be located at the effluent of
the first adsorber. A high effluent concentration from the first
adsorber indicates that this adsorber is reaching its capacity. The first adsorber is then taken off-line, and the second
adsorber is shifted to be the first adsorber. Consequently, the
capacity of both adsorbers can be fully utilized, and the compliance requirements can also be met. If there are two parallel
streams of adsorbers, one stream can always be taken off-line
for regeneration or maintenance, and the continuous operation of the system is secured.

7.2.5 COC Removal Rate by an Activated-Carbon Adsorber

The COC removal rate by a GAC adsorber ($R_{removal}$) can be calculated by
using the following formula:

$$R_{removal} = (G_{in} - G_{out})Q \tag{7.5}$$

In practical applications, the effluent concentration (G_{out}) is kept below the
discharge limit, which is often very low. Therefore, for a factor of safety, the
term of G_{out} can be deleted from Equation (7.5) in design. The mass removal
rate is then the same as the mass loading rate ($R_{loading}$):

$$R_{removal} \approx R_{loading} = (G_{in})Q \tag{7.6}$$

The mass loading rate is nothing but the multiplication product of the air
flow rate and the COC concentration. As mentioned in Chapter 2, the contaminant concentration in the air is often expressed in ppmV or ppbV. In
the mass loading rate calculation, the concentration has to be converted into
mass concentration units as:

$$1\,ppmV = \frac{MW}{22.4}\ [mg/m^3]\quad at\ 0°C$$

$$= \frac{MW}{24.05}\ [mg/m^3]\quad at\ 20°C \tag{7.7}$$

$$= \frac{MW}{24.5}\ [mg/m^3]\quad at\ 25°C$$

or

$$1 \, ppmV = \frac{MW}{359} \times 10^{-6} \quad [lb/ft^3] \quad at \ 32°F$$

$$= \frac{MW}{385} \times 10^{-6} \quad [lb/ft^3] \quad at \ 68°F \tag{7.8}$$

$$= \frac{MW}{392} \times 10^{-6} \quad [lb/ft^3] \quad at \ 77°F$$

where MW is the molecular weight of the compound.

Example 7.3: Determine the Mass Removal Rate by the GAC Adsorbers

Referring to the remediation project described in Example 7.2, the discharge limit for xylene is 100 ppbV. Determine the mass removal rate by the two 55-gal GAC units.

Solution:

(a) Use Equation (7.8) to convert the ppmV concentration to lb/ft³:
   Molecular weight of xylene $(C_6H_4(CH_3)_2)$ = 12 × 8 + 1 × 10 = 106

$$1 \, ppmV = \frac{106}{392} \times 10^{-6} = 0.27 \times 10^{-6} \, lb/ft^3 \quad at \ 77°F$$

   800 ppmV = (800)(0.27 × 10⁻⁶) = 2.16 × 10⁻⁴ lb/ft³

(b) Use Equation (7.6) to determine the mass removal rate:

$$R_{removal} \approx (G_{in})Q = (2.16 \times 10^{-4} \, lb/ft^3)(200 \, ft^3/min)$$

$$= 0.65 \, lb/min = 93 \, lb/day$$

7.2.6 Change-Out (or Regeneration) Frequency

Once the activated carbon reaches its capacity, it should be regenerated or disposed of. The time interval between two regenerations or the expected service life of a fresh batch of GAC can be found by dividing the capacity of GAC with the COC removal rate ($R_{removal}$) as:

$$T = \frac{M_{removal}}{R_{removal}} \tag{7.9}$$

Example 7.4: Determine the Change-Out (or Regeneration)
Frequency for GAC Adsorbers

Referring to the remediation project described in Example 7.3, the discharge limit for xylene is 100 ppbV. Determine the service life of the two 55-gal GAC units.

Solution:

As shown in Example 7.2, the amount of xylene that each drum can retain before being exhausted is 28.6 lbs. Use Eq. 7.9 to determine the service life of two drums:

$$T = \frac{M_{removal}}{R_{removal}} = \frac{(2)(28.6 \text{ lb})}{0.65 \text{ lb / min}} = 88 \text{ min} < 1.5 \text{ h}$$

Discussion:

1. Although two drums in parallel can provide a sufficient cross-sectional area for adequate air flow velocity, the relatively high contaminant concentration makes the service life of the two 55-gal drums unacceptably short.

2. A 55-gal activated-carbon drum normally costs several hundred dollars. In this example, two drums last less than 90 min. The labor and disposal costs should also be added, and it makes this option prohibitive. A GAC system with on-site regeneration or other treatment alternatives should be considered.

7.2.7 Amount of Carbon Required (On-Site Regeneration)

If the COC concentration of the air stream is high, a GAC system with on-site regeneration capability would become a more attractive option. The amount of GAC required for on-site regeneration depends on the mass loading, the adsorption capacity of GAC, the design service time between two regenerations, and the ratio between the number of GAC units/beds in regeneration cycle and the number of GAC units/beds in adsorption cycle. It can be determined by using the following formula [1]:

$$M_{GAC} = \frac{R_{removal}T_{ad}}{q}\left[1 + \frac{N_{des}}{N_{ad}}\right] \qquad (7.10)$$

where
M_{GAC} = total amount of GAC required
T_{ad} = adsorption time between two consecutive regenerations
N_{ad} = number of GAC beds in the adsorption phase
N_{des} = number of GAC beds in the regeneration (desorption) phase

**Example 7.5: Determine the Amount of GAC Required
                for On-Site Regeneration**

Referring to the remediation project described in Example 7.3, an on-site regeneration GAC is proposed to deal with the high COC loading. The system consists of three adsorbers. Two of the three adsorbers are in the adsorption cycle, and the other one is in the regeneration cycle. The adsorption cycle time is six hours. Determine the amount of GAC required for this system.

> **Solution:**
> The total amount of GAC required in all three adsorbers can be determined by using Equation (7.10) as:
>
> $$M_{GAC} = \frac{R_{removal}T_{ad}}{q}\left[1+\frac{N_{des}}{N_{ad}}\right] = \frac{(0.65 \text{ lb/min})(360 \text{ min})}{(0.193 \text{ lb/lb})}\left[1+\frac{1}{2}\right] = 1,818 \text{ lb}$$
>
> So, a total of 1,818 pounds of GAC is required (606 pounds in each bed).

7.3 Thermal Oxidation

Thermal processes are also commonly used to treat VOC-laden air. Thermal oxidation, catalytic oxidation, and internal combustion (IC) engines are popular thermal processes for these applications. The key components of thermal treatment system design are the three Ts, which are combustion temperature, residence time (also called *dwell time*), and turbulence. They affect the size of a reactor and its destruction efficiency. For example, to achieve good thermal destruction, the VOC-laden air should be held inside a thermal oxidizer for a sufficient residence time (normally 0.3–1.0 s) at a high temperature, at least 100°F above the auto-ignition temperatures of the COCs in the VOC-laden gas stream. In addition, sufficient turbulence must be maintained in the oxidizer to ensure good mixing and complete combustion of the COCs. Other important parameters to be considered include heating value of the influent and the requirements of auxiliary fuel and supplementary air.

Discussion on the combustion basics for thermal oxidation will be presented here, and it is essentially applicable to other thermal processes.

7.3.1 Air Flow Rate versus Temperature

The volumetric air flow rate is commonly expressed in ft³/min in the US customary system, i.e., cubic feet per minute (cfm). Since the volumetric flow rate of an air stream is a function of temperature and the air stream undergoes zones of different temperatures in a thermal process, the air flow rate is further shown as actual cfm (acfm) or standard cfm (scfm). The unit of acfm refers to the volumetric flow rate under the actual temperature, while scfm is the flow rate at the standard conditions. The standard conditions are the basis for comparison. Unfortunately, the definition of the standard conditions is not universal. For USEPA, the standard conditions are $T = 77°F$ (25°C) and $P = 1$ atm. However, they are 60°F (15.56°C) and 1 atm for the South Coast Air Quality Management Districts (SCAQMD) in southern California. In addition, 32°F (0°C) or 68°F (20°C) are also commonly used in technical articles as the temperature for the standard conditions. One should follow the regulatory requirements and use the appropriate reference temperature for a specific project. A standard temperature of 77°F is used in this chapter, unless otherwise specified.

Conversions between acfm and scfm for a given air stream can be easily made using the following formula, which assumes that the Ideal Gas Law is valid:

$$\frac{Q_{actual\ @\ temperature\ T,\ in\ acfm}}{Q_{standard,\ in\ scfm}} = \frac{460 + T\ (\text{in }°F)}{460 + 77} \tag{7.11}$$

where T is the actual temperature in °F, and the addition of 460 is to convert the temperature from degree Fahrenheit to degree Rankine. It should be noted that if the temperatures are expressed in degree Celsius, Equation (7.12) can be used for the conversions between acfm and scfm. (The addition of 273 is to convert the temperature from degree Celsius to degree Kelvin.):

$$\frac{Q_{actual\ @\ temperature\ T,\ in\ acfm}}{Q_{standard,\ in\ scfm}} = \frac{273 + T\ (\text{in }°C)}{273 + 25} \tag{7.12}$$

Example 7.6: Conversion between the Actual and Standard Air Flow Rates

A thermal oxidizer was used to treat the off-gas from a soil-venting process. To achieve the required removal efficiency, the oxidizer was operated at 1,400°F. The actual flow rate at the exit of the oxidizer was 550 ft^3/min. What would be the exit flow rate expressed in scfm? The temperature of the effluent air from the final discharge stack was 200°F, and the diameter of the final stack was 4 in. Determine the air flow velocity from the discharge stack.

Solution:

(a) Use Equation (7.11) to convert acfm to scfm as:

$$\frac{Q_{actual\ @\ temperature\ T,\ in\ acfm}}{Q_{standard,\ in\ scfm}} = \frac{460 + T}{460 + 77} = \frac{460 + 1400}{460 + 77} = \frac{550}{Q_{standard,\ in\ scfm}}$$

So, $Q = 158.8$ scfm

(b)   Use Equation (7.11) to determine the flow rate from the stack:

$$\frac{Q_{actual\ @\ temperature\ T,\ in\ acfm}}{Q_{standard,\ in\ scfm}} = \frac{460 + T}{460 + 77} = \frac{460 + 200}{460 + 77}$$

$$= \frac{Q_{actual\ @\ temperature\ T,\ in\ acfm}}{158.8}$$

So, $Q = 195.2$ acfm @ 200°F

The discharge velocity, $v = Q/A = Q \div (\pi r^2)$

$= 195.2 \text{ ft}^3/\text{min} \div [\pi(2/12)^2 \text{ ft}^2]$

$= 2{,}240 \text{ ft/min}$

Discussion:

If the actual flow rate at one temperature is known, it can be used to determine the flow rate at another temperature by using the following formula:

$$\frac{Q_{actual\ @\ T_1}}{Q_{actual\ @\ T_2}} = \frac{460 + T_1}{460 + T_2} \tag{7.13}$$

The stack flow rate in this example can be directly determined by using the exit flow rate from the oxidizer as:

$$\frac{Q_{actual\ @\ T_1}}{Q_{actual\ @\ T_2}} = \frac{460 + T_1}{460 + T_2} = \frac{550}{Q_{actual\ @\ 200°F}} = \frac{460 + 1400}{460 + 200}$$

Thus, Q_{actual} @ 200°F = 192.2 acfm

7.3.2 Heating Value of an Air Stream

Organic compounds generally contain heating values. These organic compounds can also serve as energy sources for combustion. The higher the organic concentration in a waste stream, the higher the heat content is and the lower the requirement of auxiliary fuel would be. If the heating value of a compound is not available, the following Dulong's formula can be used:

$$\text{Heating value (in Btu/lb)} = 145.4\ C + 620\left(H - \frac{O}{8}\right) + 41\ S \tag{7.14}$$

where C, H, O, and S are the percentages by weight of these elements in the compound. Equation (7.14) can also be used to estimate the heating value of a solid waste. The heating value of an air stream containing organics can be determined by:

Heating value of an air stream containing VOCs (in Btu/scf)

$$= \text{VOC's heating value (in Btu/lb)} \times \text{Mass concentration of the VOC (lb/scf)}$$
$$(7.15)$$

We can divide the heating value of a waste air stream in Btu/scf by the density of the air to obtain the heating value in Btu/lb.

Heating value of an air stream containing VOCs (in Btu/lb)

$$= \text{Heating value (in Btu/scf)} \div \text{Density of the air stream (lb/scf)} \qquad (7.16)$$

The density of an air stream under the standard condition can be found as:

$$\text{Density of an air stream (in lb/scf)} = \frac{\text{Molecular weight}}{392} \qquad (7.17)$$

Since air consists mainly of 21% oxygen (MW = 32) and 79% nitrogen (MW = 28), people normally use 29 as the molecular weight of the air. Consequently, the density of the air is 0.0739 lb/scf (= 29/392). This value can also be used for VOC-laden air, provided the VOC concentrations are not extremely high.

Example 7.7: Estimate the Heating Value of an Air Stream

Referring to the remediation project described in Example 7.1, a thermal oxidizer is also considered to treat the off-gas. Estimate the heating value of the air stream that contains 800 ppmV of xylene.

Solution:

(a) Use the Dulong's formula (Equation 7.14) to estimate the heating value of pure xylene:

Molecular weight of xylene $(C_6H_4(CH_3)_2) = 12 \times 8 + 1 \times 10 = 106$

Weight percentage of $C = (12 \times 8) \div 106 = 90.57\%$

Weight percentage of $H = (1 \times 10) \div 106 = 9.43\%$

$$\text{Heating value (in Btu/lb)} = 145.4\,C + 620\left(H - \frac{O}{8}\right) + 41\,S$$

$$= 145.4(90.57) + 620\left(9.43 - \frac{0}{8}\right) + 41(0)$$

$$= 19{,}015$$

(b) To determine the heat content of the air containing 800 ppmV xylene, we have to determine the mass concentration of xylene in the air first (which has been previously determined in Example 7.3):

$$800 \text{ ppmV of xylene} = (800)(0.27 \times 10^{-6})$$
$$= 2.16 \times 10^{-4} \text{ lb xylene/ft}^3 \text{ air}$$

Use Equation (7.15) to determine the heating value of the off-gas:

Heating value (in Btu/scf)

$$= 19{,}015 \text{ Btu/lb} \times (2.16 \times 10^{-4} \text{ lb/scf}) = 4.11 \text{ Btu/scf}$$

(c) Use Equation (7.16) to convert the heating value into Btu/lb:

Heating value of an air stream containing xylenes (in Btu/lb)

$$= 4.11 \text{ Btu/scf} \div 0.0739 \text{ lb/scf} = 55.6 \text{ Btu/lb}$$

Discussion:

1. The heating value of xylene calculated from the Dulong's formula, 19,015 Btu/lb, is essentially the same as that found in the literature, 18,650 Btu/lb.

2. The weight percentage of C is 90.57%, and a value of 90.57, not 0.9057, should be used in the Dulong's formula.

7.3.3 Dilution Air

Some waste air streams contain enough organic compounds to sustain burning (i.e., no auxiliary fuel is required, which means cost savings). That is why direct incineration is favorable for treating air with high organic concentrations. However, for hazardous air pollutant streams, the concentration of flammable vapors to a thermal incinerator is generally limited to 25% of the lower explosive limit (LEL), imposed by insurance companies for safety concerns. Vapor concentrations up to 40% to 50% of the LEL may be permissible if on-line monitoring of VOC concentrations and automatic process control and shutdown are employed. Table 7.2 lists the LELs and upper explosive limits (UELs) of some combustible compounds in air.

When the off-gas has VOC content larger than 25% percent of the LEL (i.e., in most of the initial stages of the SVE-based cleanup projects), dilution air must be used to lower the COC concentration to below 25% of its LEL prior to incineration [3]. The 25% LEL corresponds to a heat content of 176 Btu/lb or 13 Btu/scf in most cases.

TABLE 7.2

The LEL and UEL of Some Organic Compounds in Air

Compounds	LEL (% volume)	UEL (% volume)
Methane	5.0	15.0
Ethane	3.0	3.0
Propane	2.1	9.5
n-Butane	1.8	8.4
n-Pentane	1.4	7.8
n-Hexane	1.2	7.4
n-Heptane	1.05	6.7
n-Octane	0.95	3.2
Ethylene	2.7	36
Propylene	2.4	11
1,3 Butadiene	2.0	12
Benzene	1.3	7.0
Toluene	1.2	7.1
Ethyl benzene	1.0	6.7
Xylenes	1.1	6.4
Methyl alcohol	6.7	36
Dimethyl ether	3.4	27
Acetaldehyde	4.0	36
Methyl ethyl ketone	1.9	10

Source: [1].

Example 7.8: Determine the Heating Value of an Air Stream at 25% of Its LEL

An off-gas stream contains a high level of benzene. The heating value of benzene is 18,210 Btu/lb. Determine the heating value of an off-gas stream that corresponds to 25% of its LEL.

Solution:

(a) From Table 7.2, the 100% LEL of benzene in air is 1.3% by volume.

$$\text{Then, } 25\% \text{ of LEL} = (25\%)(1.3\%) = 0.325\% \text{ by volume}$$
$$= 3{,}250 \text{ ppmV}$$

Molecular weight of benzene $(C_6H_6) = 12 \times 6 + 1 \times 6 = 78$

Use Equation (7.8) to convert ppmV to lb/ft³:

$$1 \text{ ppmV} = \frac{78}{392} \times 10^{-6} = 0.199 \times 10^{-6} \text{ lb/ft}^3 \quad \text{at } 77°F$$

$$3{,}250 \text{ ppmV} = (3{,}250)(0.199 \times 10^{-6}) = 6.47 \times 10^{-4} \text{ lb/ft}^3$$

(b)   Use Equation (7.15) to determine the heating value of the off-gas:

Heating value (in Btu/scf)

$$= 18{,}210 \text{ Btu/lb} \times (6.47 \times 10^{-4} \text{ lb/scf}) = 11.8 \text{ Btu/scf}$$

(c)   Use Equation (7.16) to convert the heating value into Btu/lb:

Heating value of an air stream containing benzene (in Btu/lb)

$$= 11.8 \text{ Btu/scf} \div 0.0739 \text{ lb/scf} = 160 \text{ Btu/lb}$$

Discussion:

The calculated heating value, 11.8 Btu/scf or 160 Btu/lb, is similar to the value of 13 Btu/scf or 176 Btu/lb for 25% LELs of typical VOCs.

When dilution is required, the volumetric flow rate of the dilution air can be found as [1]:

$$Q_{dilution} = \left[\frac{H_w}{H_i} - 1 \right] Q_w \tag{7.18}$$

where
$Q_{dilution}$ = required dilution air, scfm
Q_w = waste air stream to be treated, scfm
H_w = heat content of the waste air stream, Btu/scf (or Btu/lb)
H_i = heat content of the desired influent entering the treatment system, Btu/scf (or Btu/lb)

Example 7.9: Determine the Requirement of the Dilution Air

An off-gas stream ($Q = 200$ scfm) is to be treated by direct incineration. The heating value of the off-gas is 300 Btu/lb. The insurance policy limits the COC concentration in the influent air to the thermal oxidizer to ≤25% of its LEL. Determine the required dilution air flow rate.

Solution:

Use 176 Btu/lb as the heating value that corresponds to 25% LEL. The dilution air flow rate can be determined by using Equation (7.18) as:

$$Q_{dilution} = \left[\frac{H_w}{H_i} - 1 \right] Q_w = \left[\frac{300}{176} - 1 \right] (200) = 141 \text{ scfm}$$

7.3.4 Auxiliary Air

If the waste air stream has a low oxygen content (below 13% to 16%), then auxiliary air would also be used to raise the oxygen level to ensure flame stability of the burner. If the exact composition of the waste air stream is known, one can determine the stoichiometric amount of air (oxygen) for complete combustion. In general practices, excess air is added to ensure complete combustion. The following example illustrates how to determine the stoichiometric amount of air and excess air for combusting a landfill gas.

Example 7.10: Determine the Stoichiometric Air and Excess Air for Combusting Landfill Gas

A landfill gas stream ($Q = 200$ scfm) is to be treated by an incinerator. The landfill gas is composed of 60% by volume CH_4 and 40% CO_2. The gas is to be burned with 20% excess air at 1,800°F. Determine (1) the stoichiometric amount of air required, (2) the auxiliary air required, (3) the total influent flow rate to the incinerator, and (4) the total effluent flow rate from the incinerator.

Solution:

(a)   The influent flow rate of methane = (60%)(200 scfm) = 120 cfm

The influent flow rate of carbon dioxide = (40%)(200 scfm) = 80 cfm

The reaction for complete combustion of methane is:

$$CH_4 + 2O_2 \rightarrow CO_2 + 2H_2O$$

The stoichiometric requirement of oxygen

= (120 scfm)(2 moles of O_2 per mole of CH_4)

= 240 scfm

The stoichiometric requirement of air

= (oxygen flow rate) ÷ (oxygen content in air)

= (240 scfm) ÷ (21%) = 1,140 scfm

(b)   The total auxiliary air = (1 + 20%)(1,140 scfm) = 1,368 cfm

The flow rate of nitrogen in the auxiliary air

= (79%)(1,368) = 1,080 scfm

(c)   The total influent flow rate

= 120 (methane) + 80 (carbon dioxide) + 1,368 (air)

= 1,568 scfm

(d)   The flow rate of oxygen in the effluent = (20%)(240) = 48 scfm

The flow rate of nitrogen in the effluent

= The flow rate of nitrogen in the influent = 1,080 scfm

The flow rate of carbon dioxide in the effluent

= carbon dioxide in the landfill gas + carbon dioxide produced from combustion

= 80 + 120 (CH_4:CO_2 = 1:1) = 200 scfm

The flow rate of water vapor in the effluent

= water vapor produced from combustion (CH_4:H_2O = 1:2)

= (2)(120) = 240 scfm

The total effluent flow rate = 48 + 1,080 + 200 + 240 = 1,568 scfm

Discussion:

1.   The following table summarizes the flow rate of each component in this process:

	CH_4	O_2	N_2	CO_2	H_2O
Influent (scfm)	120	2(120)(1.2) = 288	1,080	80	0
Effluent (scfm)	0	288 − 240 = 48	1,080	80 + 120 = 200	240

2.   The flow rates of the total influent and total effluent are the same at 1,568 scfm.

7.3.5 Supplementary Fuel Requirements

The VOC concentration of the off-gas from soil/groundwater remediation can be very low and insufficient to support combustion. If that is the case, auxiliary fuel would be needed. The following equation can be used to determine the requirement of supplementary fuel [1]:

$$Q_{sf} = \frac{D_w Q_w [C_p (1.1 T_c - T_{he} - 0.1 T_r) - H_w]}{D_{sf} [H_{sf} - 1.1 C_p (T_c - T_r)]} \qquad (7.19)$$

where

Q_{sf} = flow rate of the supplementary fuel, scfm

D_w = density of the waste air stream, lb/scf (usually 0.0739 lb/scf)

D_{sf} = density of the supplementary fuel, lb/scf (0.0408 lb/scf for methane)

T_c = combustion temperature, °F

T_{he} = temperature of the waste air stream after the heat exchanger, °F

T_r = reference temperature, 77°F

C_p = mean heat capacity of air between T_c and T_r

H_w = heat content of the waste air stream, Btu/lb

H_{sf} = heating value of supplementary fuel, Btu/lb (21,600 Btu/lb for methane)

If the temperature of the waste air stream after the heat exchanger (T_{he}) is not specified, use the following equation to calculate T_{he} (Note: the heat exchanger is to recuperate the heat from the exhaust of the oxidizer to heat up the influent waste air stream):

$$T_{he} = \left(\frac{HR}{100}\right)T_c + \left[1 - \frac{HR}{100}\right]T_w \qquad (7.20)$$

where

HR = heat recovery in the heat exchanger, % (If no other information is available, a value of 70% may be assumed.)

T_w = temperature of the waste air stream before entering the heat exchanger, °F

In Equation (7.20), T_{he} is the temperature of waste air stream after the heat exchanger. (If no heat exchangers are employed to recuperate the heat, then $T_{he} = T_w$.) The C_p value can be obtained from Figure 7.1.

Example 7.11: Determine the Supplementary Fuel Requirements

Referring to the remediation project described in Example 7.7, an off-gas stream ($Q = 200$ scfm) containing 800 ppmV of xylene is to be treated by a thermal oxidizer with a recuperative heat exchanger. The combustion temperature is set at 1,800 °F. Determine the flow rate of methane as the supplementary fuel, if required.

Solution:

(a) Assuming that the heat recovery is 70% and the temperature of the waste air from the venting well is 65°F, the temperature of the waste air after the heat exchanger, T_{he}, can be found from Equation (7.20) as:

$$T_{he} = \left(\frac{HR}{100}\right)T_c + \left[1 - \frac{HR}{100}\right]T_w$$

$$= \left(\frac{70}{100}\right)(1,800) + \left[1 - \frac{70}{100}\right](65) = 1,280°F$$

(b) The average specific heat can be read from Figure 7.1 as 0.0266 Btu/lb-°F at 1,800°F.

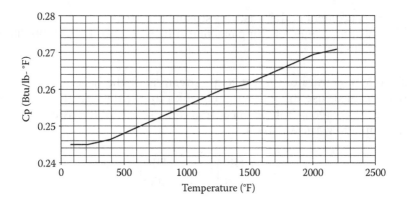

FIGURE 7.1
Specific heat of air versus temperature.

(c) The heat content of the waste gas is 55.6 Btu/lb, as determined in Example 7.7.

(d) The flow rate of the supplementary fuel can be estimated by using Equation (7.19) as:

$$Q_{sf} = \frac{D_w Q_w [C_p (1.1 T_c - T_{he} - 0.1 T_r) - H_w]}{D_{sf} [H_{sf} - 1.1 C_p (T_c - T_r)]}$$

$$= \frac{(0.0739)(200)\{(0.266)[1.1(1,800) - 1,280 - 0.1(77)] - 55.6\}}{(0.0408)\{21,600 - 1.1[(0.266)(1,800 - 77)]\}}$$

$$= 2.21 \text{ scfm}$$

7.3.6 Volume of the Combustion Chamber

The total influent to an incinerator is the sum of the waste air, dilution air (and/or the auxiliary air), and the supplementary fuel, and it can be determined by the following equation:

$$Q_{inf} = Q_w + Q_d + Q_{sf} \qquad (7.21)$$

where Q_{inf} = the total influent flow rate, scfm.

In most cases, one can assume that the flow rate of the combined gas stream, Q_{inf}, entering the combustion chamber is approximately equal to the flue gas leaving the combustion chamber at standard conditions, Q_{fg}. The volume change across the incineration chamber, due to combustion of VOC and supplementary fuel, is assumed to be small. This is especially true for dilute VOC streams from soil/groundwater remediation.

TABLE 7.3

Typical Thermal Incinerator System Design Values

Required Destruction Efficiency (%)	Nonhalogenated Compounds		Halogenated Compounds	
	Combustion Temperature (°F)	Residence Time (sec)	Combustion Temperature (°F)	Residence Time (sec)
98	1,600	0.75	2,000	1.0
99	1,800	0.75	2,000	1.0

Source: [1].

The flue gas flow rate of actual conditions can be determined from Equation (7.11) or from Equation (7.22):

$$Q_{fg,a} = Q_{fg}\left[\frac{T_c + 460}{77 + 460}\right] = Q_{fg}\left[\frac{T_c + 460}{537}\right] \qquad (7.22)$$

where $Q_{fg,a}$ is the actual flue gas flow rate in acfm.

The volume of the combustion chamber, V_c, is determined from the residence time, τ (in sec), and $Q_{fg,a}$ by using Equation (7.23):

$$V_c = \left[\left(\frac{Q_{fg,a}}{60}\right)\tau\right] \times 1.05 \qquad (7.23)$$

The equation is nothing but "residence time = volume ÷ flow rate." The factor of 1.05 is a safety factor, which is an industrial practice to account for minor fluctuations in the flow rate. Table 7.3 tabulates the typical thermal incinerator system design values.

Example 7.12: Determine the Size of the Thermal Incinerator

Referring to the remediation project described in Example 7.11, an off-gas stream ($Q = 200$ scfm) containing 800 ppmV of xylenes is to be treated by a thermal oxidizer with a recuperative heat exchanger. The combustion temperature is set at 1,800°F to achieve a destruction efficiency of 99% or higher. Determine the size of the thermal incinerator.

Solution:

(a) Use Equation (7.21) to determine the flue gas flow rate at standard conditions:

$$Q_{fg} \approx Q_{inf} = Q_w + Q_d + Q_{sf} = 200 + 0 + 2.21 = 202.2 \text{ scfm}$$

(b) Use Equation (7.22) to determine the flue gas flow rate at actual conditions:

$$Q_{fg,a} = Q_{fg}\left[\frac{T_c + 460}{537}\right] = (202.2)\left[\frac{1,800 + 460}{537}\right] = 851 \text{ acfm}$$

(c) From Table 7.3, the required residence time is 1 sec. Use Equation (7.23) to determine the size of the combustion chamber as:

$$V_c = \left[\left(\frac{Q_{fg,a}}{60}\right)\tau\right] \times 1.05 = \left[\left(\frac{202.2}{60}\right)(1)\right] \times 1.05 = 3.5 \text{ ft}^3$$

7.4 Catalytic Incineration

Catalytic incineration, also known as catalytic oxidation, is another commonly applied combustion technology for treating VOC-laden air. With the presence of a precious or base metal catalyst, the combustion temperature is normally between 600°F and 1,200°F, which is lower than that of the direct thermal incineration systems.

For catalytic oxidation, the three Ts (temperature, residence time, and turbulence) are still the important design parameters. In addition, the type of catalyst has a significant effect on the system performance and cost.

7.4.1 Dilution Air

The concentration of flammable vapors to a catalytic incinerator is generally limited to 10 Btu/scf or 135 Btu/lb (equivalent to 20% LEL for most VOCs), which is lower than that for direct incineration. This is due to the fact that higher VOC concentrations may generate too much heat upon combustion and deactivate the catalyst. Therefore, dilution air must be used to lower the COC concentration to below 20% of its LEL.

When dilution is required, the volumetric flow rate of the dilution air can be found from Equation (7.18):

$$Q_{dilution} = \left[\frac{H_w}{H_i} - 1\right]Q_w$$

Example 7.13: Determine the Requirement of the Dilution Air

Referring to the remediation project described in Example 7.8, an off-gas stream (Q = 200 scfm) containing 800 ppmV of xylenes is to be treated by a catalytic incinerator with a recuperative heat exchanger. Determine the required dilution air flow rate, if needed.

Solution:

The heating value of the off-gas has been determined as 11.6 Btu/scf or 160 Btu/lb in Example 7.8, which exceeds the 10-Btu/scf or 135-Btu/lb limit. Thus, air dilution is required, and the dilution air flow rate can be determined by using Equation (7.18) as:

$$Q_{dilution} = \left[\frac{H_w}{H_i} - 1\right]Q_w = \left[\frac{160}{135} - 1\right](200) = 37 \text{ scfm}$$

Discussion:

For the same off-gas, 800 ppmV of xylenes, air dilution is required for catalytic incineration, but not required for direct incineration.

7.4.2 Supplementary Heat Requirements

For catalytic incineration of off-gases from soil/groundwater remediation, supplementary heat is often provided by electrical heaters. If natural gas is used, use Equation (7.19) to determine the supplementary fuel flow rate. Before calculating the supplementary heat requirement, the following two equations should be used to estimate the temperature of the flue gas, T_{out}, which can achieve the desired destruction efficiency without damaging the catalyst. The T_{out} can be estimated with the temperature of the waste gas after the heat exchanger and before the catalyst bed, T_{in}, and the heat content of the gas:

$$T_{out} = T_{in} + 50H_w \tag{7.24}$$

On the other hand, the equation can be modified to determine the required influent temperature to achieve a desired temperature in the catalyst bed:

$$T_{in} = T_{out} - 50H_w \tag{7.25}$$

where H_w is the heat content of the waste air stream in Btu/scf only. These two equations assume a 50°F temperature increase for every 1 Btu/scf of heat content in the influent air to the catalyst bed.

Example 7.14: Estimate the Temperature of the Catalyst Bed

Referring to the remediation project described in Example 7.13, an off-gas stream ($Q = 200$ scfm) containing 800 ppmV of xylenes is to be treated by a catalytic incinerator with a recuperative heat exchanger. After the heat exchanger, the temperature of the diluted waste gas is 550°F. Estimate the temperature of the catalyst bed.

Solution:

After air dilution, heat content of the diluted waste gas is 10 Btu/scf. Use Equation (7.24) to estimate the temperature of the catalyst bed:

$$T_{out} = T_{in} + 50H_w = 550 + (50)(10) = 1,050°F$$

Discussion:

The calculated temperature, 1,050°F, falls in the typical temperature range for catalyst beds (1,000°F–1,200°F).

7.4.3 Volume of the Catalyst Bed

The total influent to a catalyst bed is the sum of the waste air, dilution air (and/or the auxiliary air), and the supplementary fuel, and it can be determined from Equation (7.21):

$$Q_{inf} = Q_w + Q_d + Q_{sf}$$

In most cases, one can assume that the flow rate of the combined gas stream, Q_{inf}, entering the catalyst is approximately equal to the flue gas leaving the catalyst at standard conditions, Q_{fg}. The flue gas flow rate of actual conditions can be determined from Equation (7.22):

$$Q_{fg,a} = Q_{fg}\left[\frac{T_c + 460}{77 + 460}\right] = Q_{fg}\left[\frac{T_c + 460}{537}\right]$$

Because of the short residence time in the catalyst bed, space velocity is commonly used to relate the volumetric air flow rate and the volume of the catalyst bed. The space velocity is defined as the volumetric flow rate of the VOC-laden air entering the catalyst bed divided by the volume of the catalyst bed. It is the inverse of residence time. Table 7.4 provides the typical design parameters for catalytic incinerators. It should be noted here that the flow rate used in the space velocity calculation is based on the influent gas flow rate at standard conditions, not that of the catalyst bed or the bed effluent.

TABLE 7.4

Typical Design Parameters for Catalytic Incineration

Desired Destruction Efficiency (%)	Temperature at Catalyst Bed Inlet (°F)	Temperature at Catalyst Bed Outlet (°F)	Space Velocity (h⁻¹)	
			Base Metal	Precious Metal
95	600	1,000–1,200	10,000–15,000	30,000–40,000

Source: [1].

The size of the catalyst can be determined by:

$$V_{cat} = \frac{60Q_{inf}}{SV}$$ (7.26)

where
V_{cat} = volume of the catalyst bed, ft^3
Q_{inf} = total influent flow rate to the catalyst bed, scfm
SV = space velocity, h^{-1}

Example 7.15: Determine the Size of the Catalyst Bed

Referring to the remediation project described in Example 7.13, an off-gas stream (Q = 200 scfm) containing 800 ppmV of xylenes is to be treated by a catalytic incinerator with a recuperative heat exchanger. The design space velocity is 12,000 h^{-1}. Determine the size of the catalyst bed.

Solution:
(a) Use Equation (7.21) to determine the flue gas flow rate at standard conditions:

$$Q_{fg} \approx Q_{inf} = Q_w + Q_d + Q_{sf} = 200 + 37 + 0 = 237 \text{ scfm}$$

(b) With a space velocity of 12,000 h^{-1}, use Equation (7.26) to determine the size of the catalyst bed:

$$V_{cat} = \frac{60Q_{inf}}{SV} = \frac{(60)(237)}{12,000} = 1.2 \text{ ft}^3$$

Discussion:
The size of the catalyst bed, 1.2 ft^3, is smaller than the volume of the combustion chamber for direct incineration, 3.5 ft^3.

7.5 Internal Combustion Engines

An internal combustion (IC) engine of an automobile or truck can be modified and incorporated into a system to treat VOC-laden air. The IC engine is used as a thermal incinerator, and the physical difference between the IC engine units and the thermal incinerators is mainly in the geometry of the combustion chamber.

Sizing an IC engine device is based on the volumetric flow rate of the VOC-laden air to be treated. One vendor reports that their IC engine unit can handle up to 80 scfm of VOC-laden air, while the other reports that their unit can accommodate 100 to 200 scfm of influent gas (depending on the VOC concentrations) for every 300 in.^{3} of engine capacity [2]. Conservatively speaking, a typical IC engine should not handle more than 100 cfm of VOC-laden air. For a higher flow rate, a treatment system with a few IC engines in parallel would be needed.

Example 7.16: Determine the Number of IC Engines Needed

Referring to the remediation project described in Example 7.13, an off-gas stream ($Q = 200$ scfm) containing 800 ppmV of xylenes is to be treated by IC engines. Determine the number of IC engines needed for this project.

Solution:

The average off-gas flow rate is 200 scfm, and a typical IC engine can only handle 100 scfm as the maximum. Therefore, a minimum of two IC engines in parallel should be used in this project.

7.6 Soil Beds/Biofilters

In biofiltration, the VOC-laden air is vented through a biologically active soil medium where VOCs are biodegraded. The temperature and moisture of the air stream and biofilter bed are critical in design considerations.

Biofiltration is cost effective for large-volume air streams with relatively low concentrations (<1,000 ppmV as methane). Maximum influent VOC concentrations have been found to be 3,000–5,000 mg/m^3. For optimum efficiency, the waste air stream should be at 20°C–40°C and 95% relative humidity. The filter material should be maintained at 40%–60% moisture by weight and a pH between 7 and 8. Typical biofilter systems have been designed to treat 1,000–150,000 m^3/h waste air, with the systems having 10–2,000 m^2 of filter media. The typical depth of biofilter media is 3–4 ft [2]. The typical surface loading rate is 100 m^3/h of waste air stream per m^2 filter cross-sectional area. The required cross-sectional area of the biofilter (A_{filter}) can be determined as:

$$A_{\text{biofilter}} = \frac{\text{Air flow rate}}{\text{Surface loading rate}} \qquad (7.27)$$

Example 7.17: Sizing Biofilters for Off-Gas Treatment

Referring to the remediation project described in Example 7.13, an off-gas stream ($Q = 200$ scfm) containing 800 ppmV of xylenes is to be treated by biofilters. Determine the size of the biofilters needed for this project.

Solution:

(a) The off-gas contains 800 ppmV of xylenes, which is equivalent to 6,400 ppmV as methane (each xylene molecule contains eight carbon atoms). This is beyond the typical range of <1,000 ppmV as methane. The maximum influent VOC concentrations of 3,000–4,000 mg/m³ have been reported in the literature. Although the xylenes concentration in this case (800 ppmV of xylenes = 3,460 mg/m³) falls within the range, a dilution of off-gas would be a conservative approach. The optimal influent concentration should be determined from a pilot study. In this example, let us dilute the off-gas four times; therefore, the influent flow rate to the biofilter becomes 800 scfm.

(b) The typical surface loading rate is 100 m³/h of waste air stream per m² filter cross-sectional area. Let us convert 800 cfm to m³/h as:

$$Q = 800 \text{ ft}^3/\text{min} = (800 \text{ ft}^3/\text{min})(60 \text{ min/h})(0.0283 \text{ m}^3/\text{ft}^3)$$
$$= 1,360 \text{ m}^3/\text{h}$$

Use Equation (7.27) to determine the required cross-sectional area as:

$$A_{biofilter} = \frac{\text{Air flow rate}}{\text{Surface loading rate}} = \frac{1,360 \text{ m}^3/\text{h}}{100 \text{ m}^3/\text{h/m}^2}$$

$$= 13.6 \text{ m}^2 = 146 \text{ ft}^2$$

Use a typical value of 4 ft as the depth of the biofilter.

Discussion:

If the biofilter is constructed in a cylindrical shape, the diameter of the biofilter would be around 14 ft.

References

1. USEPA. 1991. Control technologies for hazardous air pollutants. EPA/6254/6-91/014. Washington, DC: Office of Research and Development, US Environmental Protection Agency.
2. USEPA. 1992. Control of air emissions for superfund sites. EPA/624/R-92/012. Washington, DC: Office of Research and Development, US Environmental Protection Agency.
3. USEPA. 2006. Off-gas treatment technologies for soil vapor extraction systems: State of the practice. EPA/542/R-05/028. Washington, DC: Office of Superfund Remediation and Technology Innovation, Office of Solid Waste and Emergency Response, US Environmental Protection Agency.

Index